LAW AND THE VISIBLE

A Volume in the Series

THE AMHERST SERIES IN LAW, JURISPRUDENCE, AND SOCIAL THOUGHT
Edited by
Austin Sarat, Lawrence Douglas, and Martha Merrill Umphrey

LAW AND THE VISIBLE

Edited by
Austin Sarat
Lawrence Douglas
Martha Merrill Umphrey

University of Massachusetts Press
Amherst and Boston

ISBN 978-1-62534-586-8 (paper); 587-5 (hardcover)

Designed by Jack Harrison
Set in Scala
Printed and bound by Books International, Inc.

Cover design by Frank Gutbrod
Cover photo by Mateusz Chodakowski, *Two Gray CcTV Cameras on Wall*, 2017. Courtesy Pexels

Library of Congress Cataloging-in-Publication Data
Names: Sarat, Austin, editor. | Douglas, Lawrence, editor. | Umphrey,
Martha Merrill, editor.
Title: Law and the visible / edited by Austin Sarat, Lawrence Douglas,
Martha Merrill Umphrey, University of Massachusetts Press.
Description: Amherst : University of Massachusetts Press, [2021] | Series:
The Amherst series in law, jurisprudence, and social thought | Includes
bibliographical references and index.
Identifiers: LCCN 2020053341 (print) | LCCN 2020053342 (ebook) | ISBN
9781625345868 (paperback) | ISBN 9781625345875 (hardcover) | ISBN
9781613768426 (ebook) | ISBN 9781613768433 (ebook)
Subjects: LCSH: Electronic evidence—United States. | Digital video—United
States. | Wearable video devices in police work—United States. | Video
recordings—Law and legislation—United States. | Audio-visual
materials—Law and legislation—United States.
Classification: LCC KF8947.5 .L39 2021 (print) | LCC KF8947.5 (ebook) |
DDC 345.73/064—dc23
LC record available at https://lccn.loc.gov/2020053341
LC ebook record available at https://lccn.loc.gov/2020053342

British Library Cataloguing-in-Publication Data
A catalog record for this book is available from the British Library.

To Ben (A.S.)

Contents

Acknowledgments

We are grateful to our Amherst College colleagues Michaela Brangan, David Delaney, Mona Oraby, and Adam Sitze for their intellectual companionship, and to our students in Amherst College's Department of Law, Jurisprudence & Social Thought for their interest in the issues addressed in this book. We would also like to express our appreciation for generous financial support provided by Amherst College's Corliss Lamont Fund.

LAW AND THE VISIBLE

AN INTRODUCTION

Law and the Visible
New Visual Technologies and the Problematics of Judgment

MARTHA MERRILL UMPHREY, AUSTIN SARAT,
AND LAWRENCE DOUGLAS

In the past few decades—dating back at least to 1991, when the murky video of four police officers beating Rodney King on the side of a Los Angeles freeway was broadcast around the world—visual technologies have taken on new prominence in debates about the legitimate exercise of state police power, the responsibilities of witnesses to violence, and the uses of digital surveillance. Cell phone cameras, police dashcams, CCTV cameras, drones—such technologies seemingly are everywhere, transmitting with abandon. We all can now sit in front of a computer screen and watch police chases, beatings, arrests, drone strikes, or bystander and doorbell videos of apparently criminal activity. Recorded, livestreamed, uploaded, endlessly reviewed online and in courtrooms, events that in the past could be recounted only in some other medium by eyewitnesses now circulate widely as visual evidence for public consumption, interpretation, and dispute.

The essays in this volume take up this new landscape, inhabited by both the police and private citizens caught up in the logics and dynamics of visibility. The analyses they offer are particularly urgent in light of the events of 2020: digital images of the shocking deaths of George Floyd and Ahmaud Arbery circulated widely, energizing the Black Lives Matter movement, fueling nationwide protests, and generating a sea change in political conversations about racist policing practices.[1] Our contributors explore a number of questions emerging from that landscape. What is the relationship between technology, objectivity, and evidence? What are the dynamics of mediated witnessing, and what responsibilities flow from it? How should we understand flows of power and meaning when everyone has the capacity to deploy digital visual technologies? How are such technologies bound

1

up in particular with racialized and gendered histories and ideologies? In what ways might surveillance and visual exposure be inflected, corrected, or countered by law and by those being watched?

In addressing those questions, our contributors draw on scholarship in legal studies, visual studies, media studies, and surveillance studies.[2] They conceive of digital image technologies in two overlapping ways. One framework emerging out of surveillance studies, grounded in the work of Michel Foucault, understands digital surveillance and recording practices in terms of a logic of policing and capture. Videos offer objective access to reality as it actually occurred, from which vantage point one can derive the truth of events, individuals, and populations under surveillance. We can be confident of the knowledge the camera's gaze produces. A second and related framework more commonly found in media studies conceives of technologies of the visible as facilitating witnessing, providing evidence that attempts to transmit facts across epistemological gaps in ways that both depend upon and problematize truth-telling. Witnessing calls forth a sense of responsibility to the events witnessed rather than exerting power over their meanings.

The Foucauldian framework helps to explain why certain populations (racialized, gendered) are more often subject to surveillance and unwanted exposure, and also why activists critiquing such practices embrace the proliferation of cameras as a panacea for injustices visited upon their communities. Videos of police beatings and shootings seem unquestionably to provide evidence of brutality in the courtroom and in public. Yet often legal outcomes and public opinion do not accord with that interpretation. The second framework offers a way to discern why that is the case—why the proliferation of video technologies guarantees neither the truth value of the images they capture nor the justice they promise. Video offers an experience of a witnessing—indeed, it can itself act as witness online. But that experience must be taken up with care and can never finally determine the truths to which it attests.

From a variety of angles, our contributors explore these entangled problematics of capturing and witnessing, which are fundamentally and inextricably entangled with law. State surveillance in general, and police cameras in particular, extend the eyes of the law into the crevices of everyday life. Courtrooms provide for negotiating the meanings of images produced by both state and private actors. Legal and constitutional principles provide limits and remedies for unwarranted visual capture and public exposure. Ultimately, digital images are both powerful and problematic in law precisely because law generates those images in moments of contact between the state and the citizen, even as it is both called to make judgments about them and held accountable for them in complex ways by those whom law polices.

Valuing Images

Visibility is constituted by the relation between an object and its viewer. When the eye is mechanical, a simple old-fashioned film camera, for example, some theorists understand that relation to be unmediated. Film's receptivity to light makes it a kind of mirror for what is before the lens.[3] "The photograph," Roland Barthes says, "is literally an emanation of the referent. . . . Every photograph is a certificate of presence."[4] He continues, "The important thing is that the photograph possesses an evidential force, and that its testimony bears not on the object but on time. From a phenomenological viewpoint, in the Photograph, the power of authentication exceeds the power of representation."[5] In Barthes's view, photography attests to existence but not meaning. As evidence of something that happened, the photograph holds out the promise of a mechanical objectivity, not subjective witnessing. As Lorraine Daston and Peter Galison characterize this promise, "To be objective is to aspire to knowledge that bears no trace of the knower—knowledge unmarked by prejudice or skill, fantasy or judgment, wishing or striving. Objectivity is blind sight, seeing without inference, interpretation, or intelligence."[6] Conceived this way, the camera is a mirror of nature that communicates what is true, beyond human perspective or judgment.

In arguing that the photograph attests that something was present before the camera—that it has an indexical value[7]—Barthes makes a minimalist claim that nevertheless anchors more robust fantasies of photographic referentiality. In the domain of the juridical, the fantasy of such referentiality has a powerful epistemological allure because it both anchors and substitutes for the vagaries of subjective human judgment.[8] The 2007 US Supreme Court ruling in *Scott v. Harris* provides an iconic example of this allure. In this case, the dashcams in four police cruisers recorded a car chase, a collision, and the frightening and spectacular result that rendered Harris a quadriplegic.[9] He sued, alleging that Scott caused the collision by using excessive force and thereby violated the Fourth Amendment's prohibition against unreasonable seizures. Scott appealed a denial of summary judgment (that is, a judgment without a full trial, which occurs when no material facts are in dispute) to the Supreme Court, which ruled in his favor. The dashcam videos proved pivotal in helping to secure Scott's legal victory. "The videotape," Justice Scalia wrote for the majority, "quite clearly contradicts the version of the story told by respondent [Harris]. . . . We are happy to allow the videotape to speak for itself. . . . Far from being the cautious and controlled driver the lower court depicts, what we see on the video more closely resembles a Hollywood-style car chase of the most frightening sort, placing police officers and innocent bystanders alike at great risk of serious injury."[10]

To Justice Scalia, the videotape's images both possess and extend Barthes's "evidential force": they capture the light and shadow of a high-speed car chase as a thing that the camera indexically recorded. Yet, in Scalia's view, they call forth a legal judgment that attaches seemingly irrefutable meaning to the recording. His opinion reveals the ideological trajectory of that meaning-making, gliding without friction from assertions about mechanical objectivity (the video speaks for itself) to a very specific interpretive framing—the Hollywood car chase, full of adrenaline, good guys and bad guys, and the inevitable capture and containment of danger. For Scalia, authentication becomes representation, and the images' evidentiary value becomes irrefutable. Justice Stevens leaps on that elision in his dissent. The tape, he argues, "actually confirms, rather than contradicts, the lower courts' appraisal of the factual questions at issue. . . . This is hardly the stuff of Hollywood. To the contrary, the video does not reveal any incidents that could even be remotely characterized as 'close calls.'"[11] He continues: "The Court [has] used its observation of the video as an excuse for replacing the rule of law with its ad hoc judgment."[12]

Stevens's critique is a reminder that, as Donna Haraway puts it, "an optics is a politics of positioning."[13] As Daston and Galison argue, while photographs wield "a powerful ideological force as the very symbol of neutral, exquisitely detailed truth," they are never fully successful in eliminating human intervention between object and representation.[14] Barthes's modest but powerful claim about the authenticating power of the photograph can easily slide into a kind of naïve realism, noted by a number of our contributors, that denies the culturally embedded codes (e.g., Hollywood chase scenes) that frame interpretations of photographic meaning. When that slide occurs, images not only attest to something's past presence but allow us to make untroubled decisions about it; indeed, we hold out hope that they function as neutral sites of judgment.

Historian and theorist John Tagg critiques this ideology as a manifestation of realism rather than reality. Realism (as a set of generic conventions that mask the constructedness of text and image) both elides the photograph's production process and extracts images from the historical circumstances that frame them, while hiding those operations from the naïve viewer. "Every photograph," Tagg argues, "is the result of specific and, in every sense, significant distortions which render its relation to any prior reality deeply problematic and raise the question of the determining level of the material apparatus and of the social practices within which photography takes place."[15] Those distortions necessarily occur both because of the technical translation needed to turn the three-dimensional object in front of the camera into a legible two-dimensional image,[16] and because the meaning ascribed to that image cannot be acontextual and innocent.

Justice Scalia's understanding of the Harris video is framed by generic conventions that cultivate sympathy for the frightening vulnerabilities of police work. The police cruiser's dashcam enables that perspective, relaying the crash from the police's perspective alone. Neither Scalia nor Stevens comments on the ways the video formally positions viewers and invites them to identify with that perspective (a dynamic that our contributor Eden Osucha compares with the identificatory structures of first-person shooter videogames). As W.J.T. Mitchell suggests, "What is at stake in the contestation of the sensible is rarely the formal question of visual perception but the social organization and control that is mediated by it. This divide between the 'police' and the people is not an incidental aspect or by-product of political power but is constitutive of it."[17]

Thus conceived, photographic practice is intimately bound up with the project of policing and a logic of capture. Drawing on Foucault's analysis of eighteenth- and nineteenth-century disciplinary apparatuses (manifested most overtly in Jeremy Bentham's designs for the panopticon, a penitentiary designed to allow a few state actors to keep many inmates under constant surveillance),[18] Tagg's analysis of late nineteenth-century photographic practices sees them as produced by "the forms and relations of power which are brought to bear on practices of representation or constitute their conditions of existence, but also the power effects which representational practices themselves engender."[19] This perspective inverts the realist association between image-production and truth-production; rather than reflecting "reality," photographic practices *produce* the "reality" that the camera captures. Meaning is ascribed to that "reality" in ways constituted by the power dynamics in play. The subject in front of the lens is someone or something already inscribed into an preconceived framework: in the surveillance video we see a "criminal," in the bodycam, a "thug," on the phone screen, a "slut."[20] As Osucha, Kelli Moore, and Carrie Rentschler in particular suggest, those meanings are inseparable from the exercise of social power, anchored by long histories of racial and gender subjugation.

And yet with the advent of new technologies of the visible, cameras just as easily can be turned against state actors. As surveillance studies scholars Haggerty and Ericson argue, the proliferation of handheld cameras now levels hierarchies of visibility, placing everyone under surveillance in a "synoptic" regime that inverts the visual dynamics of the panopticon, allowing the many to watch the few.[21] This reversal, often framed as a kind of political resistance, nevertheless remains within the visual (if not social) logic of panopticism. Viewers still understand images as capturing, in an archetypal case such as the George Floyd killing, the essential truth of police and state violence. The broad circulation of police beating and shooting videos often creates widespread public outrage, and those videos' immediate reception often indicates a powerful and understandable desire to ascribe objective evidentiary value to them. Calls for the use of dashcams and bodycams, as well as eyewitness advocacy, are grounded in the promise that images can render the exercise of power objectively visible and transparent, and hence open to reform.

As a number of our contributors note, that promise often remains unfulfilled. Images are reinterpreted, unconscious bias may structure the judgment of jurors, and officers are acquitted even in the face of widespread public condemnation. Take, for example, the 2016 shooting of Philando Castile, one of many cases in which video problematized rather than guaranteed truth. That July, police in a Minnesota suburb pulled over Castile, an African American school cafeteria worker—perhaps because of a broken taillight, perhaps because he resembled an armed robbery suspect. After a forty-second exchange captured on a police dashcam, Officer Jeronimo Yanez shot Castile in front of his girlfriend, Diamond Reynolds, and her four-year-old daughter, who were also in Castile's car. After the shots were fired, Reynolds began to livestream the bloody aftermath of the shooting and her own arrest, broadcasting shaky, fragmented images from a phone. By the following day the video had been viewed almost 2.5 million times on Facebook. The killing generated a national outcry, with protests, some violent, occurring in the ensuing weeks. It also elicited statements of concern from Governor Mark Dayton and President Barack Obama.

As media studies scholar Penelope Papailias notes of the Castile livestream, "Reynolds' act of testimony clearly demonstrates how networked mobile image technologies contribute to the emergence of new visual idioms and ethical discourses of mobile witnessing that interrupt and reshape 'cultures of record.'"[22] The power of such images is undeniable; livestreaming in particular seems to be quintessentially objective and tamper-proof, broadcasting images in real time.[23] Yet, our contributors suggest, even its truth claims can be contestable and contested.

Officials charged Yanez with second-degree manslaughter and reckless discharge of a firearm, yet he was acquitted after trial. Immediately following the verdict, officials released dashcam footage, as well as additional footage taken from inside Yanez's squad car, showing Reynolds's young daughter heartrendingly comforting her keening mother. In that fraught context, what did Reynolds's and the police's images signify? The endlessly recirculated videos offer neither the pretense of Haraway's objective, god's-eye view nor the confident truth-production of the panopticon, making impossible a definitive interpretation of the visual confusion generated by Reynolds's dramatic livestreaming, set alongside videos from two different cameras in the police cruiser.[24]

Yet if they cannot fully capture her terrifying experience in ways we might describe as "objective," nevertheless Reynolds, her livestream, and the cruiser's cameras each bear witness to the aftermath of Castile's shooting in ways that carry moral and epistemological weight. Witnessing, John Durham Peters notes, signifies both a sensory experience and the discursive act of stating one's experience for an audience that was not present at the event and yet must make some kind of judgment about it. A witness is "an observer or source possessing privileged (raw, authentic) proximity to facts. . . . What one has seen," Peters continues, "authorizes what one says: an active witness first must have been a passive one," and in that sense a "witness is the paradigm case of a *medium*: the means by which experience is supplied to others who lack the original."[25]

All the case's video footage attests to Castile's death, and to Reynolds's and her daughter's proximity and response to it. Yet even as witnesses (whether human or technological) recount the facts they have experienced, Peters argues, they must negotiate the difficult juncture between that experience and discourse. "The witness," he writes, "is authorized to speak by having been present at an occurrence. A private experience enables a public statement. But the journey from experience (the seen) into words (the said) is precarious."[26] Treating images as witnesses suggests that meaning-making takes place within a dynamic logic that recognizes instabilities between image and meaning, evidence and judgment, even witnesses, victims, and offenders.[27]

If the evidentiary truth value of Reynolds's images seems elusive and contingent, nevertheless they embody an ethics of recording, disseminating, and testifying about police violence that powerfully and convincingly grounds calls for racial justice. Carrie Rentschler would describe both Reynolds and her phone as "digital witnesses,"[28] and, following Kelly Oliver, would characterize her livestreaming as an exercise in "response-ability," the active capacity to respond to violence.[29] In Peters's terms, Reynolds's witnessing of and to Castile's death moves from passive to active the moment she begins to livestream its aftermath, signaled by shifts in the audience she addresses. "Stay with me," she says to Castile when the video begins; and then immediately she turns to her online audience. "We got pulled over for a busted tail light in the back and the police just, he's covered. He killed my boyfriend," she continues. No matter that Yanez may have had other motives; no matter that Castile has not yet stopped breathing—from the perspective of witnessing, the live images transmit a searing experience, unveiled in real time. We hear Reynolds address Yanez, and we hear him and another officer bark orders at her. She talks to Jesus, to Facebook, and to her four-year-old daughter, and her camera supplies all that experience for her viewers. The camera, sometimes on the ground, sometimes off-angle, frames her testimony by overtly displaying the conditions of its production, transmitting the shooting's aftermath in ways (in Rentschler's terms) that are responsible to her boyfriend, her daughter, and her community.

Scott v. Harris, the Castile case, the Floyd and Arbery cases, and others like them invite us to note the complexities that attend the valuing and evaluation of digital images in contemporary culture. As some of the essays below explore, such images resonate with prior images of past acts of state and institutional violence. That prehistory can inform and reinforce the viewer's sense that images captured by the camera lens transmit an unalloyed and true rendering of events—they seem to happen today as they have happened again and again before—in ways that reinforce our desire to believe that what we see is objectively referential. At the same time, our contributors argue, viewers ought to be wary of that desire, and of the correlative imputation that technology can fully replace human judgment in assessing the meaning of digital images. In contexts that range well beyond the visual capture of police violence on Black bodies, they suggest that visual images can neither replace legal processes as evidentiary grounds for judgment nor provide stable grounds for a critique of state surveillance. However complex the task may be, to do justice both law and its critics must develop the capacity to interpret, supplement, contextualize, and resist the images new visual technologies generate.

The Essays

Our contributors survey the landscape of new visual technologies from a variety of disciplinary and interdisciplinary perspectives; yet taken together they express a general skepticism about the promises of those technologies for law and its critique. Jennifer Petersen and Kelli Moore dive into the logics of technological truth-production, highlighting the cultural constructedness of even the most seemingly objective surveillance and predictive mechanisms. Petersen explores contradictions in popular and legal conceptions of evidence, proof, and objectivity as they play out in the use of video evidence in the courtroom. Drawing on Daston and Galison's history of various conceptions of objectivity, she challenges continuing faith in the neutrality and objective evidentiary value of the image. Arguing that nineteenth-century ideas about the link between photography and objectivity have been revived in contemporary faith in automation and computational processing, Petersen highlights the tension between assertions of mechanical/computational objectivity and the embodied nature of witnessing and testimony. If in the nineteenth century, objectivity was equated with restraining the observer's personal bias and judgment with a "view from nowhere," today we locate that disembodied perspective in video technologies.

Those technologies, Petersen argues, cannot fulfill their promises. In elaborating this claim, she critically examines two seemingly different kinds of video evidence: red-light cameras and eyewitness videos. Red-light cameras, which automatically photograph vehicles running red lights, create a regime of panoptic surveillance that seems to offer a kind of computational objectivity. Removing human subjectivity from the framing of an image, they replace humans with computers that do the mechanical processing and reading. Eyewitness videos go further, offering testimony not just about whether an event happened but also what it was, in some cases overriding other kinds of testimony.

Yet in both cases, Petersen argues, perspective is both constitutive of the images and unavoidable. Computer imagery uses algorithms and manipulation to produce usable evidence, and relies on misplaced trust in authorizing institutions. Video images operate through realist aesthetics of perspective, something advocacy groups that train eyewitnesses to video police encounters use to their advantage in order to develop a shared meaning and moral evaluation. Ultimately, Petersen argues, discussions of the value of video as evidence should engage with procedures, policies, and people dictating its production.

Kelli Moore, in "Eye-Tracking Techniques and Strategies of the Flesh in *The Brother from Another Planet*," offers a skeptical view of the role that new technology, and more specifically computational algorithms and machine learning, can play in reforming liberal legal regimes. In the context of a growing awareness of police violence against communities of color, Moore offers a meditation on a metaphor—the "eyes of the law"—to capture how legal spectatorship is increasingly and problematically oriented in ways that resonate with relations of enslavement. How, she asks, are "slave-like" practices implicated in questions about bystander/witness perception of police-civilian interactions, particularly in relation to juror decisions to punish?

Moore focuses on eye-tracking technologies that purport to measure perceptual bias, identifying "ethno-somatic responses" in observers watching video interactions between White and Black people. Eye-tracking machines analyze the engagement of attention and unconscious gaze patterns through algorithms and interpretation, while the participants are surreptitiously tracked. They generate rhetorically compelling descriptions of the ways different groups unconsciously perceive and punish threat; but that such data comes from (as Maurizio Lazzarato suggests) the machinic enslavement of participants generates its own specific neurological rendering of perception. Reading that theory of enslavement alongside Orlando Patterson's analysis of human slavery as social death, Moore suggests that machinic enslavement accomplishes something similar in colonizing the unconscious. She turns to cinema, and specifically three films—*A Clockwork Orange, Bladerunner,* and *The Brother from Another Planet*—to illustrate the ways film has anticipated a machine-oriented way of conceiving the eyes of the law. Ultimately, Moore argues, data generated from machine learning enhances predictive policing and forecasts perceptual bias, reproducing a racialized logic in legal spectators. When prediction and prejudice align to generate a logic of legal spectatorship, she suggests, we should be skeptical of the judgments that result.

Carrie A. Rentschler and Eden Osucha draw our attention to the ethical, political, and legal complexities of video witnessing, both human and technological, and the kinds of accountability it can and cannot generate. Rentschler, in "Mediating Responsibility: Visualizing Bystander Participation in Sexual Violence," explores the connection between digital and legal witnessing in cases involving livestreamed and recorded sexual assault. What does it mean, she asks, to bear ethical and legal responsibility as digital bystanders and witnesses, particularly in contexts that do not enforce a duty to rescue someone being injured? Bystander videos and photos can become records of complicity and culpability, but also serve as evidence to report and prosecute sexual assault. Making bystanders visible as digital witnesses, Rentschler argues, can shift and broaden our conception of responsibility—or response-ability, the capacity to respond to violence but also the ethical obligation to do so, acting on calls for help—to encompass the larger perpetrator-identified culture of participants.

Rentschler explores the connections between recording and enacting violence, and the potential complicity of those who do the recording, in two particular cases: the 2016 case in which Marina Lonina livestreamed a friend's sexual assault; and the 2013 case of Rehtaeh Parsons, who committed suicide after the online circulation of photos of her sexual assault. Lonina was charged with sexual assault even as she claimed she had livestreamed the assault as a call for help; the men involved in the Parsons rape were convicted of making and distributing child pornography but not sexual assault. In both examples, witness responsibility is enacted by making witness irresponsibility audible and visible. How, Rentschler asks, might we impose not a duty to rescue but a duty to enact response-ability—to testify after the fact and take responsibility—in such cases? Canadian law has responded by criminalizing the making and distributing of sexually exploitative images but not by conceptualizing bystanders as participating in the assaults themselves; they are guilty of the recording but not the act recorded. And yet, Rentschler concludes, in spite of the fact that bystanders occupy the role of digital witness, their responsibility remains contingent upon both their visibility in the recordings and our interpretations of them.

Eden Osucha, in her chapter "Between the Bodycam and the Black Body: The Post-Panoptic Racial Interface," interrogates the assumption that video from bodycams offers transparent access to the truth of police violence against Black and Brown communities. To understand why bodycams are ineffective as a remedy to contain police violence, she argues, we need to situate the videos they produce in particular historical and cultural contexts that frame images of Black suffering in unstable, ambivalent ways. Two surveillance systems generate what she calls the "post-panoptic" interface racializing the indexical evidentiary value of these videos: surveillance of communities of color, and an online landscape that inscribes them in a longstanding, morally and socially hierarchized visual system in which portraits can be both dignifying "honorific images" and repressive reinscriptions of the suffering Black body.

Osucha traces the emergence of the bodycam, initially embraced as a mechanism for greater justice after the 2014 shooting of Michael Brown in Ferguson, Missouri, and situates it within a powerful scopic regime inherited from the practices of Whites during and after the era of slavery. The public consumption of spectacles of Black suffering is not a new phenomenon made possible by the Internet; it dates to the slave market, the plantation overseer, and the souvenir postcard depicting lynchings. In those postcards the camera functions as a weapon of violence against the Black body, extending and sanctioning the reach of racist practices even as the images on them were also deployed by anti-racism activists as evidence of atrocity. In offering a reading of Samuel DuBose's 2015 shooting by officer Ray Tensing, Osucha argues that bodycam videos are similarly ambiguous: they mimic the perspective of first-person shooter videogames, interpolating the viewer as the shooter, and raise challenging questions about whether their images force a moral engagement with the killing of African Americans or a close identification with the killer.

Osucha argues that the circulation of such images online constructs a particular genre in which African Americans are inscribed in longstanding scripts of Black suffering that may be watched for pleasure. Such videos also delimit the structures of perception through which law sees anti-Black police violence. They individuate the events depicted rather than addressing the harms of institutionalized racism's less visible forms. The dissemination and reduplication of such images sap them of their vitality and singularity, affectively deadening them in ways that facilitate their reinterpretability. If videos from bodycams gain urgency and credibility because of the evidence of police violence they seem to offer, ultimately they become emblems of an emerging society of control in which state power is exerted in mutable, mobile, and multimodal ways against a racialized other.

Finally, Torin Monahan and Benjamin J. Goold take up two different modes of critiquing and resisting state surveillance, one visual and the other jurisprudential. Monahan, in his chapter "Visualizing the Surveillance Archive: Critical Art and the Dangers of Transparency," explores the work of a number of artists who critique institutional surveillance practices, both state and corporate, as well as the archives of everyday life such practices construct. Drawing on Foucault, Monahan frames archives as technologies of power that generate truth claims out of ostensibly neutral practices— discursive systems whose parameters set normative limits on what can be included, said, and known. Archival reason, and particularly its focus on small details, resonates with surveillance practices that amass, classify, and process data in the name of transparency and control. The artists Monahan examines critique such surveillance practices by challenging their realist pretensions with images that are blurry, easily degraded, fragmented, and multiplied into meaninglessness. They destabilize the truth claims that surveillance images promulgate. Looking for traces of institutional surveillance in the world around them, these artists construct counter-archives illuminating the partiality of the archival categories that surveillance engenders.

Even so, by issuing calls for greater state and corporate transparency, these artists paradoxically embrace the very logic they critique. If institutional impulses to gain knowledge and control through surveillance lead to abuses that in turn provoke calls for more transparency, those calls themselves ignore the politics of archival categories. The "predictable valorization of transparency" one finds in artists' photographs of underwater cables or renderings of Google street view surveillance images or postings of repetitive photos of an individual's life will not, Monahan argues, lead automatically to accountability or legal change. Such calls for greater transparency are themselves reliant on surveillance-based visuality, and its invasions of privacy, as a privileged form of knowing and governance. In the end, Monahan suggests, these counter-archives operate as variations of the liberal archive, amending but not fundamentally challenging archival reason.

Benjamin J. Goold, in "Becoming Invisible: Privacy and the Value of Anonymity," argues that one way to resist increasingly encroaching regimes of state surveillance is to develop a robust defense of a right not just to privacy but to anonymity. Privacy, Goold suggests, has been presented as an individual right, and that presentation makes it more difficult to defend when the public feels endangered. Yet privacy also has a public value, protecting the exercise of key political rights (e.g., freedom of expression, religion, and free association) and guarding against the growing threat of authoritarianism. It marks and enforces the limits of state power. As we sacrifice more and more privacy in our digital lives, Goold argues, we need protection from state surveillance in order also to protect political discourse and protest.

To that end, he embraces a right to anonymity, the right to be free from identification, as something distinct from privacy and as an end in itself. Anonymity differs from privacy insofar as it is concerned with visibility, the problem of being seen, rather than with some quality or characteristic inhering in an individual. Removing indicators of personal identification enables individual freedom and solitude, and is central to the exercise of political rights.

Goold acknowledges that anonymity can shelter and even generate antisocial behavior. Fears about crime and terrorism are real. Yet, he suggests, we may underestimate the threat posed in the current moment by government overreach. While some limits on anonymity make sense (enforced against online hate speech and harassment, for example), Goold argues that law enforcement should rely less on extensive and invasive digital surveillance and more on traditional law enforcement practices. As even liberal states increasingly wish to track dissidents, a right to anonymity forwards the project of creating physical and virtual surveillance-free spaces. Goold concludes that we need to create spaces in which we cannot be known. Ironically, we may need law's help in escaping its own gaze.

Together our contributors generate a view of law as double-voiced and ambivalent, embracing the evidentiary and testimonial value of the images that new technologies produce even as it provides, or should provide, a forum for contesting their capacity to ground judgment and generate justice. If, in the name of greater transparency, we justify deploying these technologies in the hope they can unveil a shadow world of state and private violence, we need to be fully aware that the facts and data they produce cannot substitute for or displace law's own meaning-making processes. To do otherwise would be to fall prey to the ruses of realism and the hubris of the camera's gaze. The project of justice must dwell alongside and encompass the production of knowledge through visibility, our contributors suggest, without ultimately being displaced by it.

Notes

1. On May 25, 2020, White Minneapolis police officer Derek Chauvin pinned George Floyd (accused of passing a counterfeit $20 bill at a convenience store) to the ground with a knee on his neck for over eight minutes as three other officers watched. Floyd died, and witness videos circulated online immediately, igniting months of national protests. See "How George Floyd was Killed in Police Custody," https://www.nytimes.com/2020/05/31/us/george-floyd-investigation.html. A few months earlier, on February 23, 2020, Ahmaud Arbery was chased down and shot by two White men, Gregory and Travis McMichael, in Satilla Shores, Georgia. The McMichaels said they thought Arbery resembled a suspect in recent break-ins and performed a "citizen's arrest." A video of the shooting recorded by a third White man, William Bryan, went viral two months later and generated such pressure that state authorities intervened in the stalled investigation of the McMichaels. See "What We Know about the Shooting Death of Ahmaud Arbery," https://www.nytimes.com/article/ahmaud-arbery-shooting-georgia.html; "Man Who Filmed Ahmaud Arbery's Death Is Charged with Murder," https://www.nytimes.com/2020/05/21/us/william-bryan-arrest-ahmaud-arbery.html, accessed August 1, 2020.
2. Relevant recent work in legal, media, and surveillance studies includes Jessica Silbey, "Cross-Examining Film," *University of Maryland Law Journal of Race, Religion, Gender and Class*, no. 1 (2008): 17; Jennifer L. Mnookin, "The Image of Truth: Photographic Evidence and the Power of Analogy," *Yale Journal of Law & the Humanities*, no. 1 (1998): 1; Kerstin Schankweiler, Verena Straub, and Tobias Wendl, eds., *Image Testimonies: Witnessing in Times of Social Media*, Routledge Studies in Affective Societies (Abingdon:

Routledge, 2019); Anne Wagner and Richard K. Sherwin, eds., *Law, Culture and Visual Studies* (New York: Springer, 2014); Amit Pinchevski and Paul Frosh, *Media Witnessing: Testimony in the Age of Mass Communication* (London: Palgrave Macmillan, 2009); Sharrona Pearl, ed., *Images, Ethics, Technology* (London: Routledge, 2016); essays by Neil Feigenson, Jessica Silbey, and Jennifer Mnookin in "Commentaries: Visual Possibilities/Visual Persuasions," *Law, Culture, and the Humanities* 10, no. 1 (2014); Kirstie Ball, Kevin D. Heggerty, and David Lyon, *Routledge Handbook of Surveillance Studies* (New York: Routledge, 2014); Gary Marx, "Surveillance Studies," *International Encyclopedia of the Social & Behavioral Sciences*, 2nd ed. (New York: Pergamon, 2015): 733–41.

3. Obviously, digital cameras complicate the mirror metaphor because they store pixelated images in code, an infinitely malleable medium. On the history of photographs as legal evidence, see Mnookin, "The Image of Truth."

4. Roland Barthes, *Camera Lucida: Reflections on Photography*, trans. Richard Howard (New York: Hill and Wang, 1981), 80, 87.

5. Barthes, 89.

6. Lorraine Daston and Peter Galison, *Objectivity* (New York: Zone Books, 2007), 17. They argue that "mechanical objectivity" differs from other understandings of objectivity as "truth to nature" and "trained judgment," and emerged in the latter half of the nineteenth century as science embraced a moral vision of objectivity grounded in the valorization of self-restraint and self-discipline. See also Lorraine Daston and Peter Galison, "The Image of Objectivity," *Representations* 40 (Autumn 1992): 98, where they argue that "the image, as standard bearer of objectivity, is inextricably tied to a relentless search to replace individual volition and discretion in depiction by the invariable routines of mechanical reproduction."

7. Some critics have assailed Barthes's distinction between the denotative and connotative aspects of photography as unsustainable. See, for example, Allan Sekula, "On the Invention of Photographic Meaning," in *Thinking Photography*, ed. Victor Burgin (London: Macmillan Education, 1982), 84–109. On indexicality, see Mary Ann Doane, "Indexicality: Trace and Sign: Introduction," *Differences* 18, no. 1 (2007): 1–6.

8. See Mnookin, 14–27.

9. *Scott v. Harris*, 550 US 372 (2007). The case's underlying facts: in 2001, just before 11 pm on a rainy spring night in rural Georgia, nineteen-year-old Victor Harris surged away from a police car trying to pull him over for speeding. Other police cars joined the pursuit. After a high-speed chase through red lights and around other vehicles, and a diversion into a parking lot where Harris sideswiped Deputy Timothy Scott's cruiser and once again took off, Scott ultimately rammed Harris's car at high speed to stop him. Harris crashed and was severely injured. For critiques of *Scott v. Harris*, see Naomi Mezey, "The Image Cannot Speak for Itself: Film, Summary Judgment, and Visual Literacy," *Valparaiso Law Review* 48 (Fall 2013): 1–39; Peter Brooks, "Scott v. Harris: The Supreme Court's Reality Effect," *Law & Literature* 29, no. 1 (2017): 143–49; and Dan M. Kahan, David A. Hoffman, and Donald Braman, "Whose Eyes Are You Going to Believe? *Scott v. Harris* and the Perils of Cognitive Illiberalism," *Harvard Law Review* 122, no. 3 (January 2009): 837–906.

10. *Scott*, 378–80.

11. *Scott*, 390–93.

12. *Scott*, 394.

13. Donna Haraway, *Simians, Cyborgs, and Women: The Reinvention of Nature* (London: Routledge, 1990), 193. As she puts it in another well-known essay, "The eyes have been used to signify a perverse capacity—honed to perfection in the history of science tied to militarism, capitalism, colonialism, and male supremacy—to distance the knowing subject from everybody and everything in the interests of unfettered power." This

is what she terms the "god trick" of seeing everything from nowhere: an "ideology of direct, devouring, generative, and unrestricted vision . . . the gaze that mythically inscribes all the marked bodies, that makes the unmarked category claim the power to see and not be seen, to represent while escaping representation. This gaze signifies the unmarked positions of Man and White." Donna Haraway, "Situated Knowledges: The Science Question in Feminism and the Privilege of Partial Perspective," *Feminist Studies* 14, no. 3 (Autumn 1988): 581–82.

14. Daston and Galison, "The Image of Objectivity," 98, 111.

15. John Tagg, *The Burden of Representation: Essays on Photographies and Histories* (Amherst: University of Massachusetts Press, 1988), 2. Indeed Tagg critiques Barthes as a realist himself, one that approaches photographs nostalgically rather than as material artifacts. See also Allan Sekula, "The Body and the Archive," *October* 39 (Winter 1986): 3–64.

16. As Tagg describes this process with a nineteenth-century camera, "Reflected light is gathered by a static, monocular lens of particular construction, set at a particular distance from the objects in its field of view. The projected image of these objects is focused, cropped and distorted by the flat, rectangular plate of the camera which owes its structure not to the model of the eye, but to a particular theoretical conception of the problems of representing space in two dimensions. Upon this plane, the multicoloured play of light is then fixed as a granular chemical discolouration on a translucent support which, by a comparable method, may be made to yield a positive paper print. How could all this be reduced to a phenomenological guarantee?" Tagg, 3.

17. W.J.T. Mitchell, *Picture Theory* (Chicago: University of Chicago Press, 1994), 16.

18. Michel Foucault, *Discipline and Punish: The Birth of the Prison*, trans. Alan Sheridan (New York: Vintage, 1979), 195–228.

19. Tagg, 21. Tagg argues that the evidentiary value of the photograph is bound up with a genealogy tied to "the emergence of new institutions and new practices of observation and record-keeping . . . and to the development of a network of disciplinary institutions." Tagg, 5

20. As Foucault suggests, it is "a power that insidiously objectifies those on whom it is applied." Foucault, 220. In a contemporary context, this framing of visual practices emphasizes the role of state power in policing populations—particularly, in the United States, populations of color—and is particularly relevant to critiques of state surveillance practices.

21. Kevin D. Haggerty and Richard V. Ericson, eds., *The New Politics of Surveillance and Visibility* (Toronto: University of Toronto Press, 2006), 5–6. See also David Lyon, "9/11, Synopticon, and Scopophilia: Watching and Being Watched," in Haggerty and Ericson, 41–42.

22. Penelope Papailias, "Witnessing to Survive: Selfie Videos, Live Mobile Witnessing, and Black Necropolitics," in Schankweiler, Straub, and Wendl, 109.

23. Papailias, 113.

24. On the successes and failures of video evidence in conveying the experience of the overpoliced, see Susan A. Bandes, "Video, Popular Culture, and Police Excessive Force: The Elusive Narrative of Over-Policing," *University of Chicago Legal Forum* (2018): 1–23.

25. John Durham Peters, "Witnessing," *Media, Culture & Society* 23 (2001): 709.

26. Peters, 710. As Peters notes, law recognizes this precarity in the courtroom insofar as it demands a controlled, rule-bound setting in which testimony is given under oath, subject to cross-examination. Peters, 716.

27. Sacha Simons, "Credibility in Crisis: Contradictions of Web Video Witnessing," in Schankweiler, Straub, and Wendl, 19.

28. See also Simons, 21.

29. See Kelly Oliver, *Witnessing: Beyond Recognition* (Minneapolis: University of Minnesota Press, 2001).

CHAPTER 1

Ubiquitous Video, Objectivity, and the Problem of Perspective in Digital Visual Evidence

JENNIFER PETERSEN

Cameras today seem to be everywhere: in our phones, our doorbells, on the streets, in the skies. Cameras are used by police, border patrol, and the private security industries employed by wealthy individuals as tools of surveillance. Yet cameras—especially cell phone cameras—have also been used as tools of sousveillance: tools for monitoring the actions of the powerful and holding them up for judgment. Central to both surveillance and sousveillance is the promise of photographic evidence: that photographs can show us what happened and be used to hold others to account. In particular, in cases of sousveillance and acts of witness against state violence, photographs and video are positioned as tools for demanding accountability.

Yet, to date, citizen photographs and video have been more effective as popular than as legal evidence. This is most starkly demonstrated by the repeated circulation of images of police killings of Black men in the United States in recent years. From the killing of Freddie Gray in Baltimore in 2015 to the livestreaming of the killing of Philando Castille in St. Paul in 2016 to the suffocation of George Floyd in Minneapolis in 2020, we have seen again and again evidence of police violence and impunity. These videos have been important in the formation the Black Lives Matter social movement and have made many White onlookers aware of the huge stakes of structural racism in policing.[1] Yet they have rarely been successful in holding individual officers or police forces accountable in courts of law.[2]

The failures of these examples of photographic and video evidence to bring legal judgments of wrongdoing on the part of police point to the contradictions in popular and legal conceptions of photographic evidence, proof and objectivity. Much of the journalistic and policy discourse around ubiquitous cameras has suggested that they perfect modes of witnessing

17

and evidence, based in human perception but abstracted from the biases of any individual person. Yet, in practice, ubiquitous cameras—and automated ones—do not solve problems of liberal governance so much as highlight some long-standing fractures.

In this chapter, I unpack the way that hopes for witnessing and evidence via citizen video have more to do with symbolism and history than with the actual efficacy of such automated systems. The current proliferation of visual documentation created by ubiquitous cameras (still and video) raise several issues: (1) the amplification of the long-standing "mechanical objectivity" attributed to cameras by automation or "computational objectivity"; (2) the conflation of mechanical, computational, and absolute objectivity (and the differential roots of this conflation for photographs and computation); and (3) the deep tension between notions of mechanical and computational objectivity and the embodiment of witnessing and testimony. This tension can be seen in two different poles, if not extremes, of video evidence: the use of visual surveillance to document a defined event or infraction (the most pervasive and banal example being the red-light camera) and the production of eyewitness videos (e.g., to police violence or other abuses of power). In the first, automated and ubiquitous visual recording promises to dis-articulate surveillance—and visual evidence—from any particular, embodied perspective. In the latter, the perspective of the camera and the processes of identification are discursively heightened—and the site of discussions of the validity of the visual documentation in question.

Photographic Images, Evidence, and Objectivity in the Nineteenth Century

We know, and have known long before the current spate of stories about police body cameras, that video evidence does not speak for itself—and cannot speak back to powerfully entrenched ideological-affective investments such as racism. We know, or should know, that a technological device like a camera does not work alone as a deterrent to police use of force. A camera without a governing social system—a disapproving eye, or the threat of accountability—does little. Take the recent study of police body cameras in Washington, D.C., for example. The study surprised many by finding that the body cameras did not make a notable difference on police behavior. The journalistic commentary on these findings largely focused on the technology itself and wondered why the cameras did not alter behavior.[3] But cameras without a system of accountability will do nothing to change police behavior. If one has impunity, it does not matter whether anyone is watching.

The continuing fascination on the part of policymakers, journalists, and the public with police cameras despite such evidence points back to the late nineteenth century—or rather revives and intensifies a way of talking about photographs as testimony or evidence that has conceptual roots in the nineteenth century but has been revived and intensified through early twenty-first century attachments to computing. Photographs took on a (problematic) association with objectivity though a set of encounters with science and law in the late nineteenth century. This association has been intensified in automated photography, to the extent that automation and computational processing have been recruited to claims of neutrality and objectivity. Today, automated cameras can seem doubly objective, adding a layer of modern computational objectivity to the somewhat battered notions of photographic objectivity.

The invention of the daguerreotype in 1839 and the development of more modern processes of photography in the following years presented a radically new form of representation—one that many contemporaneous commentators suggested relied not on the agency, and subjectivity, of humans but rather on nature. It was common to ascribe a sort of impersonal authorship or agency to the process of daguerreotypes and photographs in the mid-nineteenth century.[4] Samuel Morse called photography "heliography," locating the authorship of photographs in the sun—a rhetorical turn that would be picked up in some late nineteenth-century legal discourse.[5] Henry Fox Talbot championed the idea that the camera was "the pencil of nature"; and that photographs were not copies of nature, but rather "portions of nature herself."[6] Similarly, John Marey suggested that photographs produced inscriptions of nature that were not transcribed into another representational system; rather, he said, "Let nature speak for itself." Photographs, so Marey and a host of other late nineteenth century scientists suggested, merely captured the natural world and natural phenomena on their own terms, so to speak.[7]

The linguistic metaphors here are fascinating. For a medium that we think of as pictorial, producing images that operate differently from words, whether spoken or written, photographs were given language frequently in the nineteenth-century discourse. The linguistic metaphors are even more striking when we consider that the documents that photographs competed with, and in many cases came to replace, were artistic and scientific images. In art, photographs were in competition with some portraiture, engravings, and other forms of "mechanical" drawing, where the point was to accurately copy (not interpret) whether this was for the purposes of documentation or of instruction (photographs were touted as excellent teaching tools for human drawing, a baseline for interpretation).[8] In science, photographs were touted as a replacement for engravings and scientific illustrations. Their usefulness, and adoption, as Lorraine Daston and Peter Galison point out, had as much to do with the moral qualities imbued to the machine in the late nineteenth century as with their accuracy or ability to depict an object. Reviewing the adoption of photography in scientific documentation, Daston and Galison point out that many hand-drawn illustrations were more accurate, in that they provided greater detail than the photographic representations in circulation, which might have less visual detail due to vagaries of exposure and to the grayscale of film.[9] Yet, at the same time as mechanical copies, they offered no new ideas or interpretation: no human intervention. In this way, photographs, even though not always the best copy, seemed to be free from subjectivity: the theories, judgments, and biases of human recorders. Machines could not tire, err, or succumb to temptation to paint the world according to their theories or preferences (as we would say today, their biases).

> What the photograph ... offered was a path to truthful depiction of a different sort, one that led not by precision but by automation—by the exclusion of the scientist's will from the field of discourse. On this view any sacrifice of resemblance was more than justified by the immediacy of the machine-made images of nature that eliminated the meddlesome intervention of humans: authenticity before mere similarity. The search for this rendition of objective representation was a moral, as much as a technical, quest.[10]

They were paragons of a certain form of virtue that was coming to define scientific norms of looking and of objectivity: the absence, or at least restraint, of human subjectivity or will.

Daston and Galison call this notion of objectivity as the absence of human judgment "mechanical objectivity"—in which the impartiality, or lack of subjectivity, of the machine becomes an ideal type for human knowledge production. (In addition, in this discourse on mechanical objectivity, the camera as a machine was at times given a sort of supra-human agency: an ability not only to see but to speak.)[11] One of the upshots of Daston and Galison's discussion of mechanical objectivity is to point out that it is different from what came before—truth to nature—and from "absolute" objectivity as knowledge or representation that is an absolutely accurate depiction of reality; this understanding of objectivity denotes what is objectively *true*.[12]

Alluding to the aesthetic dimensions of this discussion, this absolute form of objectivity is often described using Thomas Nagel's term, the view from nowhere—a knowledge without perspective. As Daston and Galison point out, mechanical objectivity, or the creation of a copy without the distortions of human bias, does not guarantee such truth—or a view from nowhere.[13]

This is true not only of human renderings, but also of human-built technologies. Media scholars like Lisa Gitelman and Jonathan Sterne point out the ways that technologies of inscription (like the photograph and the phonograph) encapsulate contemporaneous notions of human perception and value in their very design, so human and historically contingent ways of seeing—both aesthetic and political—are literally built into the apparatus.[14] In the case of photography, this is most glaringly exemplified in the way that celluloid film (and technical standards for commercial film development) was calibrated to whiteness. Film was, by the early twentieth century, designed specifically to register White skin properly;[15] digital cameras have largely replicated this bias in their attempt to reproduce the look of celluloid (this bias of an old technology—film—was a major factor in the recent Google scandal in which a Google facial recognition program misidentified African American faces in photographs as gorillas).[16]

Yet, in the nineteenth century, especially in science, criminology, and the law (as we will see), the photograph was discussed as free from human perspective or bias. This did not, as Daston and Galison caution, mean that it was understood as a route to absolute objectivity; rather, it was understood to be a hedge against subjectivity and certain types of bias it might contain. In this sense, we can read the photograph and the mechanical objectivity it was emblematic of in the sciences as precursors to procedural approaches to objectivity in the social sciences (and journalism) in the early twentieth century, in which researchers and reporters, aware of their own biases and that they would color any act of simple observation and transcription, devised techniques and procedures for measurement and recording that attempted to restrain the observer's personal judgment and bias: from the journalistic practice of recording two sides of an issue rather than attempting to determine facticity or accuracy to the focus on quantification and techniques of weeding out personal biases, such as inter-coder reliability, in the social sciences.[17] It is important to remember that such procedures did not guarantee or equal absolute objectivity. Yet, in popular discourse at least, the various meanings of objectivity have become conflated. We confuse procedural or mechanical objectivity with absolute objectivity, understood as absolute facticity.[18] In the case of photography, the very alignment of the

photograph with nature, in which the agent behind the photograph was
not a man or machine but nature itself, may have helped in this conflation:
presenting the photographic image as the view from nature approximates
the view from nowhere that, in many twentieth-century discussions of
objectivity, became coterminous with absolute objectivity (in which nature,
or God, rather than a human, acts as the sensory register of experience and
guarantee of accuracy).[19]

Objectivity, Evidence, and Law

The status of the photograph as legal evidence inherited this confusion of
objectivities. Further, the discussion of photographs as displacing human
sensory perception and human observers created a paradox within legal
practice that troubled the status of photographs as evidence. In law as well
as science, the fact that photographs relied on the sun, or nature itself, was
initially central to the ability of photographs to show the facts of a case.
However, in the United States of the late nineteenth century, the mechanical
objectivity of photographs was not sufficient for them to be easily accepted
as evidence. Photographs presented a paradox. On the one hand, they were
understood to be accurate and faithful depictions of events, and so seemed a
perfect form of evidence. On the other hand, there was no ready category for
photographs. In the absence of an appropriate category, photographs were
treated like testimony, a form of evidence based on fallible and subjective
human perception and memory. Given the linguistic metaphors employed
in the scientific discourse, perhaps it is no surprise that photographs were
brought into courtrooms as another form of testimony.

Yet photographs were a problematic form of testimony. The truth of
testimony is usually vouchsafed by the character of the speaking witness.
Further, the structure of legal testimony, in particular the semi-public facing
nature (and punitive potential) of the courtroom provide incentives toward
truthful and accurate testimony. Lies or misrepresentations will rebound
on the reputation of the witness; legal fines or imprisonment offer more
material penalties.[20] Photographs, however, had neither character nor
structural disincentives. (In place of character or interpersonal markers
of trustworthiness, photographs had what Daston and Galison refer to as
the morality of machines.) In addition, as others have pointed out, it is not
possible to cross-examine a photograph.[21]

In the late 1860s, photographs were proposed as evidence in a number
of cases, including several involving contested signatures. Most famously,
photographic evidence was part of an 1870 case involving accusations of

forgery of a will. State-of-the-art forensic techniques were used to determine the authenticity of the signature on the will, including a probabilistic analysis (completed by philosopher, semiotician, and mathematician Charles Sanders Peirce) of the signature and photographs of authentic signatures to compare to that on the will.[22] The photographs were admitted, over the objections of the defense attorney, who called them "hearsay of the sun."[23] In this term, perhaps drawing on the linguistic metaphors in circulation regarding the scientific status of film, the lawyer suggested that photographs were inadmissible, as unsubstantiated talk, or testimony that could not be authenticated. Joel Snyder argues that this assertion, that photographs were inadmissible as evidence, relied on the notion that photographs merely repeated (in their "mechanical objectivity") a statement by the sun, turning the sun into an out of court witness.[24] What is remarkable here is the simultaneous recognition of the vagaries of photographic representation—the lawyer in the Howland case raised questions about the quality of the lens, the weather, the color of the ink, the quality of the chemicals involved, even *the skill of the operator*[25]—and the placement of these vagaries outside the boundaries of not only cross-examination, but of human action, in the hands of nature. Despite the fact that it would be possible to cross-examine the creator of the photograph—the operator whose skill was even referenced by the lawyer calling for the dismissal of photographic evidence—the lawyer was able to elide the actual creator in favor of a symbolic, rhetorical one (the sun).

This was no doubt in part simply a tactic to discredit some of the expert testimony of the other side. Yet it was not an isolated argument, and the status of the sun as a witness was brought in by both sides (a Georgia Supreme Court judge argued, in favor of photographs, that there was no more reliable witness than the sun).[26] There was a lively discussion about whether photographs were originals or copies, primary evidence or secondary evidence, superior or inferior to eyewitness testimony—with some observers of the day suggesting that all testimony was but a copy—and those that relied on human memory a poor copy at that.[27] Others suggested that while the photograph might be a record produced outside the courtroom in a manner similar to hearsay, it differed from hearsay in a significant way:

> It is wholly free from the infirmity which causes the rejection of hearsay evidence, namely, the uncertainty whether or not it is an exact repetition of what was said by him whose testimony is repeated by the witness. In the picture we have before us, at the trial, precisely what the apparatus did say. Its language is repeated to us, syllable for syllable.[28]

While a tactic, the categorization of photographs as hearsay, or uncorroborated testimony, was a legible one, which made sense within the contemporaneous discourse of mechanical objectivity, in which photographs were prized as the inscription of nature.

This points to the paradox of photographs in late nineteenth-century law. Photographs were rhetorically constructed as non-human representation—a mechanical copy, the discourse of nature, objective. But they were being used as a form of evidence whose legal epistemology is based in embodied and subjective human experience. They were both more than and less than human testimony. Assertions that photographs were hearsay argued that they were less than testimony; claims they were objective renderings elevated them above human testimony.

As Jennifer Mnookin argues, claims to photographic objectivity have often referenced absolute objectivity. Exaggerated claims of photographic objectivity seemed poised to replace the fact-finding role of judges and lawyers, if not their judgment entirely. If photographs represented a real, accurate, objective rendering of a crime, for example, there would be no need for lawyers or judges to determine the case.[29] In the face of this ambiguity, legal practitioners crafted a new category of evidence: demonstrative evidence. Demonstrative evidence, or the representation of facts or claims, acts as a visual aid to illustrate or support testimony (or other evidence). Photographs would not "speak" for themselves but rather aid and support the speech of others.[30]

Yet in practice (through popular epistemology), photographs retained a patina of truth, and operated persuasively as proof. In this way, Mnookin argues, classifying photographs as supporting evidence worked not so much to curb the persuasiveness of photographs but rather to create a "new epistemic category [of evidence] that hovered uncomfortably on the boundary between illustration and proof."[31] Thus, the judicial response to photographic evidence in the late nineteenth century produced new regimes of legal visibility premised on aesthetics and notions of objectivity.

These regimes of legal visibility persist today. They are evident in judgments like the now infamous Supreme Court ruling (discussed further below) that overturned a lower court's finding of facts based on video evidence that, the majority argued, was contradicted by the facts so plainly on view in a dash-cam video. The status of video as evidence today, however, is variegated. In some instances, photographs in a very real way replace human witnesses and testimony, becoming in some ways a more powerful source of proof (assigned the status of an impartial view from nowhere, eliding questions of perspective and aesthetics, as well as interpretation). Other examples highlight the role of perspective in video evidence.

In order to explore some of the ways that photographs come to speak as evidence (or fail to do so) in the early twenty-first century, I briefly outline two case studies: red-light cameras and citizen eyewitness video, used to monitor police or state abuses of power. The points of difference between these two cases help illuminate the parameters of, and tensions around, visual evidence in the current moment. Red-light cameras produce still images (photographs), are automated, are extensions of policing, and have a very fixed and impersonal point of view. On the other hand, eyewitness videos of the sort I focus on here are moving images, are created by specific people for a variety of reasons, are often used to document police interactions (police the police) or produce testimony,[32] and tend to have embedded and complicated points of view. Perhaps the most important difference between the two examples, however, is in what they purport to document and the stakes of the documentation. The red-light camera is used to document whether or not a specific, defined event (running a red light) occurred or not. Further, the consequences of this documentation are relatively minor: a ticket and a record of poor judgment on one's driving and insurance record. Eyewitness videos have a much more complex task and often much higher stakes; they frequently seek not just to document whether or not an event happened, but what sort of event happened. That is, most often they do not aim to define whether an altercation between an individual and the police happened, but rather to define and establish that interaction as an abuse of power.[33]

Red-Light Cameras and Computational Objectivity

In the current moment, the old, if complicated, story of photographs and objectivity is being layered with a new one, of computational objectivity. Automated visual recording technologies promise not just the mechanical objectivity of the photograph, but also computational objectivity. If photographs in the late nineteenth century addressed problems of selective memory and bias about what was in the frame—questions of fidelity—then automated cameras promise to go beyond this, removing human subjectivity from the framing itself. In many ways the discourse (and legal reasoning) around automated photography today recapitulates that of the late nineteenth century, though today the promise of objectivity resides not in nature replacing human agency and bias, but in computers and numerical processing doing so.

Red-light cameras offer an instructive example, as they are common, and regularly function as evidence for traffic citations. The cameras literally stand in for human police presence, freeing up police to tend to higher priority issues while also extending the power of the police, operating as a form of panoptic surveillance in a way that clearly marked police presence cannot. They operate simply and in a relatively uncontested way. Traffic cameras are mounted on poles near an intersection. A triggering system goes off when a car enters the intersection after the light has turned red, signaling the camera(s) to photograph the intersection.[34] (The systems usually have two trigger points, one at the entry to the intersection and one partway through it, and take two photographs to document the speed the car was traveling.) Police then use the license plate to issue a ticket for running the light, sending it to the address to which the car is registered.[35]

Challenges to the legitimacy of such systems reveal the workings of these automated cameras as evidence. In response to these challenges, judges must clarify the validity and legitimacy of automated photographs as evidence. An example comes from a 2009 case in California. A driver in the city of Inglewood, part of the Los Angeles metropolitan area, fought the validity of the use of automated camera photos after she was issued a $436 traffic ticket.[36] Lawyers for the driver challenged the red-light camera photos as evidence, claiming they were hearsay, because they had been taken by a machine and had no one to vouch for them (as had those intervening against photographic evidence in the late nineteenth century). There was enough ambiguity for the case to be appealed up to the California Supreme Court, which ruled that the automated photographs were an appropriate form of evidence (*People v. Carmen Goldsmith*, 2012):

> Existing law, known as the hearsay rule, provides that, at a hearing, evidence of a statement that was made other than by a witness while testifying at the hearing and that is offered to prove the truth of the matter stated is inadmissible, subject to specified exceptions. Existing law provides that a printed representation of computer information, a computer program, or images stored on a video or digital medium is presumed to be an accurate representation of the computer information, computer program, or images that it purports to represent.[37]

The ruling built on an earlier one, *People v. Martinez* (2010), that held the actual data stored on the computer is presumed to be accurate, that digital records were not hearsay and that "the admission of computer records does not require foundational testimony showing their accuracy and reliability."[38]

In this dispute over the admissibility of automated photographs as evidence, there were many echoes of the late nineteenth-century history traced above. The case hinged on the validity and trustworthiness of automated, electronic systems of surveillance—and whether the recordings produced in such systems rose to the level of evidence. The litigant had claimed that the photographs were hearsay, not admissible as evidence. The justices on the California Supreme Court, in response, professed a deep faith in computational systems and automated photographs, arguing that they were more faithful and accurate renderings than human testimony.[39] Computer-generated data or images could not be hearsay because there were no persons involved. They reasoned that hearsay was defined as statements out of court and that only persons, not machines, could make statements; further, it was impossible to cross-examine a machine—much like the sun. Because there were no humans involved, the photographs could be considered neither statements nor hearsay. Yet, in their own words, such photographs could "purport" to represent; in this turn of phrase, and the judgment within which it resides, the people and agencies who made the photographs represent the scene are erased in favor of the isolated automated system.

These may be expeditious arguments in order to admit a common category of evidence without excessive cost or hassle. Yet they reveal assumptions about sources of unreliability in human testimony and the perspective of the machine. It is possible to label computer outputs, from Google search results to the recommendations of services like Netflix, Cortana, or Alexa, as statements. Indeed, in some other areas of law, judges and justices have presumed such outputs (i.e., search results) to be the statements protected by free speech law.[40] That the justices here presumed that computer outputs are not statements was, I would argue, a product of the context: the type of evidence (that documents whether or not a specific event occurred) and the way they see hearsay and human witness testimony. Human statements in this context suggest partial perspective, frailty of memory, and the vagaries of interpretation: they are partial, low-fidelity rendering of an event. Computer-generated photographs, on the other hand, are perfect copies of the event, high-fidelity renderings of the real. They required no cross examination to ferret out inconsistencies or for purposes of authentication.[41] There was no question of whether the cameras rendered the full picture—in part because of the type of event being recorded and in part, I suggest, because of faith in the non-technical systems and procedures defining camera placement, processing, and data storage.

As in the case of photographs, the working of a machine helped to render this vantage point invisible, or at least impartial. Yet, the fact that the machine is in this case a computer, brings in important differences. The objectivity of the photograph was premised on not only the absence of a human viewpoint from the reproduction of the event photographed, but an absence of a human operator, and at times even the absence of a human interpreter. Automated photographs compound mechanical objectivity with computational objectivity.

In photographic inscription, the promise of objectivity resided in the mechanical inscription of images. In computational objectivity, the promise of objectivity is extended to the mechanical reading—or processing and organizing, sense-making—of the images as well. It is similar to mechanical objectivity in that the route to objectivity is removal of problematic human judgment.[42] I would add that a key point of difference is what is opposed to human bias, that which anchors the objectivity of the image in each articulation of objectivity. Whereas in the late nineteenth century it was nature, in the early twenty-first it is the number, or numeracy.

As current work on computation and algorithms point out again and again, numeracy does not in fact guarantee a view from nowhere, and what we sometimes think of or talk about as purely numerical processes are in fact social ones. Just as technologies of inscription encode subjective human social theories and biases, so too do processing technologies. Computer circuits, software, and equations embody human perspectives, in the term's double sense as aesthetics and epistemology: chips draw on short cuts to produce images optimized for the human eye and equations and software describe or help bring about social purposes.[43] And the procedures employed by computers to process images and other data reflect the work of a variety of people, from those who design and implement the systems to those who process the data.[44] While the red-light camera photos involve minimal processing, this is not the case for all "automated" video evidence. As Kelly Gates demonstrates, even impersonal surveillance video (CCTV footage) must be highly produced in order to operate as evidence. Gates documents the training and work of technicians charged with producing video evidence, which includes intervention in image files, enlargement of frames, and averaging of images. She highlights the tension between the human intervention and judgment required to make the video speak in court and the status of the video as the product of objective computer systems: "The fact that the production of usable evidence often *requires* the manipulation of images would seem to present an inherent challenge to the status of those images as evidence."[45] Gates argues that a faith in computers and their moral status—computational objectivity—papers over this contradiction.

Even beyond these technical problems, or ways that human perspective is written into computational systems (software *and* hardware), it's not clear that computer or automated processing gets around the problem of bodies and their differential social authority. As with photographs in nineteenth-century courts, the ability of even automated images to speak in court relies not only on notions of computational objectivity, but also on authorizing institutions—and the many people who work within them. The reasons to put faith in the output of red-light cameras have much to do with the processes and work that go into defining the parameters of their operation and local implementation. This involves the work of technicians, elected officials, and police officers, as well as processes of legislation (creation of laws and policies governing camera use) and of local traffic law enforcement.

While the authorizing work of these institutions and individuals is effaced in the discussion of automation and invocation of computational objectivity, I suggest that much of the nonproblematic status of red-light cameras in court resides in trust of these people and institutions. The judges in the cases analyzed above do not question the framing of the images (the perspective) or the ability of the stationary red-light images to document a moving violation because they trust not only the algorithms used to trigger the cameras and the equations used to assess speed as objective in themselves, but also the authorizing institutions and the people who prescribe the placement of cameras and those who create the equations.

Not everyone has the same perspective. For those predisposed to see traffic laws and enforcement as a public good, there is often a presumption of trust. For those who see traffic laws and enforcement as a vehicle for municipal revenue (corruption), there is skepticism about every decision: camera placement, the equations used, data processing and storage.[46] In this sense, the ability of the photographs to appear as objective evidence, beyond the purview of human intervention or judgment (indeed, beyond irrationality, as the California Supreme Court argued), appears to rest in part on the identity of the institutional body behind the camera. I suggest that questions of perspective are ignored or evaluated to be irrelevant, not only because of the nature of the event being documented, but also because of the authority behind the camera.

Making Eyewitness Videos Speak:
Inscribing Perspective

Red-light cameras document a simple, well specified, event: cars passing through an intersection during a red light. The question of guilt or innocence hinges on a yes or no answer: did the event happen? Photographs in this case promise to demonstrate the truth of the allegations. They make a strong truth claim about a narrow event. And, for the most part, they seem to tell us only about the event (did X happen or not?) and not about the nature of the event, or its meaning. Eyewitness videos, however, are engaged in a much more complex set of questions. They often provide evidence not only of whether an event happened, but also of what the event was.

This is the central issue in videos of police violence. The steady stream of bystander videos of police killings of Black men in recent years has failed to convince many courts (or grand juries) of police misconduct. The videos, which seem to so many to speak for themselves, fail to provide sufficient testimony to bring criminal charges in most instances. These videos follow, in many ways, the path of the first widely circulated video of police violence against a Black man, that of police officers beating Rodney King in Los Angeles in 1991. The video failed to convince a jury of police misconduct, highlighting the way that lawyers and jurors could read a very different scene than did the videographers, and much of the public.[47] No one disputed the fact that Rodney King was arrested, or that the police used batons on him. What was disputed, and what the jury determined that the video did not demonstrate, was whether this beating was a response to a threat (and thus deemed a legitimate use of force) or an abuse of force.[48] Given the complexities of what these videos document, one might expect that evidentiary claims and uses of such video are more modest than those of red-light cameras. This is not necessarily the case. To date, some legal practitioners and courts have treated such videos as having not only mechanical objectivity, but strong objectivity.

The objectivity of video is perhaps most explicitly discussed in appeals cases, in which a lower court decision is appealed on grounds of video evidence. There is currently a split in legal reasoning on this, with some state and circuit courts defer to lower courts' determinations of facts in such cases while others opt to reinterpret the evidence, or ignore the lower court's finding of fact, in light of video. In the latter approach, judges or justices may ignore or override other (oral) testimony, as well as other judge's interpretation of the facts, based on the presumed objectivity of video evidence.[49] To override other testimony and, even more so, the judgment of other jurists suggests that these judges and justices are presuming not only mechanical (or mechanical and computational) objectivity but also strong objectivity—or, as in the discussions of late nineteenth-century photographs, conflating the two. This seems to have been the logic in the famous *Scott v. Harris* case, in which the Supreme Court overturned the verdicts of two lower courts that the actions of the police in a high-speech car chase (ramming into the rear of the fleeing suspect's car, causing him to lose control and crash and ultimately rendering him paraplegic) could be considered excessive force. The Supreme Court's ruling, that there could be no dispute about the facts of the case, meant that the officer could claim immunity from excessive force charges. The idea that there could be no dispute rested on the majority's characterization of the video evidence as a brute statement of fact—the facticity of the video, per the majority, contradicted the lower court's findings and authorized the Supreme Court to overturn their rulings. That is, they relied on the video over other testimony and over the judgment of the lower courts.[50] This was, as a number of commentators have argued, a major shift in the treatment of summary judgment and unusual in the lack of deference to earlier decisions. The idea that video evidence offered an impartial view from nowhere (and objective account) was the rationale offered to legitimate these aberrations from the norm.[51] Rather than treating video as one more source of embodied evidence (with a particular perspective), the Court relied on notions of video as a privileged window onto reality.

Scott v. Harris became famous with a wider public both for the justices' reliance on video evidence and their publication of the video evidence on the Supreme Court website to demonstrate the incontrovertibility of the majority opinion—in Scalia's words, to allow the video to "speak for itself."[52] In effect, the majority presented its reading of the video within its opinion as factual via reference to the video, confusing interpretation of images with the subject of those images. In semiotic terminology, this is not only a confusion of sign with referent, but also of an interpretation of the sign with the referent.[53] The assertion that the video spoke for itself did meet with some skepticism. Justice Stevens, in a dissent, noted that there were multiple interpretations of the video and critiqued the majority's willingness to engage in a fact-finding role best left to a jury[54] (the fact that the majority were willing to enter into the fray about the facts of the case was a product of their belief that they were not engaging in interpretation but only reasonably responding to a record of reality). And, famously, a group of law professors did an experiment in interpretation of the video, finding that the interpretation of what the video showed—a reasonable action by a law enforcement officer or recklessness and excessive force—varied with race, political ideology, and education.[55] One interpretation of these results is that what the different viewers perceived in the video varied with their likelihood of identifying with the eye of the police dashcam, a likelihood that depended in part on social position in relation to police.

One conclusion to draw from these high-profile cases is that images become, or operate as, evidence not only through technical standards like resolution, but also through the aesthetics and politics of perspective. Perspective and identification, I would like to suggest, are central to complex evidence that seeks to document what type of event happened: questions like whether a use of force was appropriate or reasonable, whether an action in a high-speed car chase was reckless, or whether a police shooting was a "reasonable" reaction to a perceived threat. These questions are not just of facts but of facts and moral judgments together.

This is not lost on advocacy groups looking to use video as evidence. These groups are keen observers of video evidence who have, through trial and error, defined their own guidelines for the best practices for producing video to provide evidence for a particular version or interpretation of events (moral evaluations). While these are not formal rules of evidence, they are rules of thumb developed in consultation with legal professionals and out of successes and failures in deploying video as evidence in a variety of types of courts.[56] One example of most effective practices in producing eyewitness video as legal (and public) evidence comes from the Los Angeles Community Action Network (LACAN). LACAN is a grassroots group advocating for homeless residents of the city; one of its aims is to ensure that homeless people's civil rights are respected by police. LACAN advocates have developed tools and best practices for videoing encounters with the police for use as evidence in lawsuits. As Forrest Stuart shows, LACAN has developed these tools with an eye to the need to craft evidence, or to the fact that video cannot necessarily speak for itself. Based in part on the lessons of the Rodney King verdict, and on trial and error, LACAN members emphasize particular indicators of realism and authenticity and harness the authority of the police to interpret events. When training new members on how to produce eyewitness video, the group emphasizes filming an entire interaction from start to finish so there is no necessity to edit. Editing, the group found, introduces questions of whether they were manufacturing an argument (or overstating the case).[57] Unedited footage, on the other hand, is not so open to such questions. The long shot is commonly understood as an index of realism.[58] The ability of the long shot to register unvarnished reality derives from the late nineteenth-century discussions of photography as the absence of human judgment or subjectivity (mechanical objectivity), in that the lack of editing implies a lack of human intervention.

Yet this indicator of realism is not enough. It could show that the events in dispute really happened (were not the product of editing) but not what kind of event it was. That is, it is not sufficient to render the video evidence of a violation of the civil liberties. As Stuart details, the best way to make eyewitness video work as evidence against the police in legal complaints, LACAN found, is to harness the voices and authority of the police in service of their argument. That is, they make the video speak through the words of the police officer on the scene. In a sense, the eyewitness videographers work to trick the officers into testifying against themselves. While filming, they ask the police to explain themselves, drawing officers in to conversation on LACAN's terms or frames, or get them to describe their actions in terms of a legal violation or to admit ignorance of laws. Such techniques provide testimony from credible witnesses (the police) in service of the argument (and legal actions) of disempowered homeless people.[59]

Similarly, WITNESS, an international organization dedicated to using photography and video to expose and prosecute human rights abuses around the globe, provides suggestions on how to create video that has a greater chance of acting as evidence—of speaking in court. WITNESS is a US-based non-profit group that partners with organizations around the world to use video to document human rights abuses. The organization's slogan is "See it. Film it. Change it"; it is one of the preeminent sources of information about producing video as evidence of such abuses. While WITNESS engages in the creation of videos to publicize human rights abuses (to provide popular evidence), much of its focus is on producing videos that can function as evidence in courts of law.[60] It produces training manuals to help individuals on the ground produce such videos and has worked to construct itself as a legitimating institution—a institutional "body" behind the camera.[61]

While the organization website has multiple guides, including guides to the different types of evidence that video can provide (and cannot), the video field guides offer the most detailed instructions on how to produce a video as evidence. In addition to details about metadata that can tie the video to a particular recording device and time, as well as demonstrate that the recording has not been altered, the materials offer aesthetic advice geared toward establishing the point of view or location of the camera eye. The guides stress not just documenting action in front of you as it unfolds (as the meaning of these events will probably not be evident to viewers), but filming a scene so that an outsider can understand it. They step out how to do so, drawing on techniques for establishing spatial and temporal continuity common in narrative film. The guides advise beginning, like any good cinematic scene, with an establishing shot capturing the location of the camera before zooming into a medium range shot to capture the action and context and then continuing with close-up shots of faces for purposes of identification. This advice urges videographers to take excessive care in documenting their perspective, the point of view of the camera, drawing on classical cinematic techniques of spatial and temporal continuity (establishing shot, medium shot, close-up) to anchor the events documented on video in a particular time and place and in correct temporal order.[62]

The guides detail techniques for spatial continuity, such as beginning with an establishing shot, moving in to a medium shot to show more detail or a close-up to focus attention on a telling detail. These shots should not be darting around, disorienting the viewer, but should be stable shots of at least ten seconds, each from different angles and perspectives on the action. In this way, the field guides instruct would-be eyewitnesses to consider their audience, and to make their films intelligible and persuasive to them. One of the key means by which they suggest videographers make their footage clear and legible is through techniques of visual storytelling and filmic conventions (common to commercial fiction films, documentary, as well as television news).[63] The registers of authenticity provided by raw, on-the-scenes footage are downplayed for the registers of Hollywood realism, founded on techniques of spatial and temporal continuity.

The advice on producing eyewitness video detailed in these manuals emphasizes the need to tell a particular story, to make the events legible to those watching the video, and to help them see in the documented events the same meaning or moral evaluation. To do so, they draw on aesthetic techniques of realism, continuity, and immersion honed in Hollywood cinema, documentary filmmaking, and television news production. These techniques make the videos speak within existing frames of visual and narrative interpretation. They are, furthermore, recruited to invite lawyers, judges, and juries to see events from a particular point of view. It is a point of view that, in the codes of Hollywood as well as classic documentary cinema, is often third-person narration, a view from above, if not nowhere.

One of the key obstacles to the ability of eyewitness videos to stand as evidence in court, then, may be perspective and identification. They do not show altercations from the point of view of police officers or other avatars of authority. They show them from the point of view of victims, or bystanders. The lesson of advocacy groups like LACAN is that these perspectives lack the authority needed to make the videos speak. One way of interpreting this is that when judicial identification fails, such videos lose their ability to appear as objective evidence.[64]

Conclusion

Evelyn Fox Keller details how the development of linear perspective in the representational arts provided a model of objectivity and created a paradox in that objectivity. Linear perspective provided a system of rules for establishing a neutral point of view, a view from nowhere. Yet this view was in fact premised on a particular, literal point of view: paintings and drawings were created to address an imagined viewer looking at the scene from outside (from a distance, as if through a window). This viewpoint, Keller argues, inaugurated a new form of "veridicality" that was also a paradox, a veridicality that "locate[d] in the vantage point of a particular somewhere at least the tacit promise of a view from nowhere."[65] This paradox was imported into the legal discussion of visual evidence in the late nineteenth century. It is still present in our discussions of visual evidence in the twenty-first. It is present in the way that automated photography (and video) is admitted as evidence in the law in California courts. If in the nineteenth century, photographs were hearsay of the sun, today automated photographs are not hearsay because there appear to be no bodies, but machines and trusted systems. This paradox is evident in the contemporary contradiction between legal discourse on the impartiality of video evidence and the work of advocates to inscribe perspective within video to make it work as evidence.

The twenty-first-century cases examined here show very different types of visual evidence used to answer very different questions (namely, did an event occur versus what kind of event was it?). Yet, I have suggested in both that the bodies behind the camera and perspective, or how we are invited to look at an event, from what angle and in what relation to the action depicted, continue to be incredibly important. In eyewitness videos of complex events, which seek to characterize events as crimes or abuses of power, the work of perspective is front and center, on display. It is also lurking in the background in more mundane discussions of video evidence (e.g., red-light cameras). In the California Supreme Court's discussion of automated photo evidence, veridicality was located in the impartiality of computers and computational systems. I argue that this was to some extent a displacement. The justices' belief in the photos, their ruling that such systems did not need to be authenticated in order to be used as evidence, rested on the impartiality of such machines (and the impossibility of machinic irrationality). This is reasonable up to a point. Yet, at least part of the reason that the justices find it unnecessary to query the validity of red-light cameras has to do with the authority of those placing and operating the cameras—the institutional body behind the lens.

Legal discussions of automated visual evidence should engage with the procedures, policies, and people that dictate the production of automated visual evidence. These, I think, are the real stakes of ubiquitous, automated visual recording as legal evidence. Notions of computational objectivity cover over or distract from such processes in a similar manner as did notions of nature as the photographer in late nineteenth-century discussions of photographic evidence.

Similarly, the fact that advocacy groups emphasize hijacking the voices of authority and/or aesthetic techniques for producing identification as among the most efficacious means of producing video evidence suggest that practices of evaluating video evidence in court telescope discussions of aesthetics and meaning (as well as social power and authority) onto the technology. In doing so, courts run the risk of conflating identification and interpretation with factual evidence. It is too easy for those who are watching to ignore perspective when it aligns with their own—or that of a trusted institution. One outcome of this partiality of perspective is a tendency to downplay or distrust eyewitness videos from marginalized perspectives, undercutting the evidentiary potential of this videos (and the power of sousveillance).

In both eyewitness videos and red-light cameras, questions of perspective and authority to speak are transferred onto the camera and social questions are treated as technical ones. Rather than treating eyewitness videos and automated cameras as totally different animals, we might do better to analyze the production, aesthetics, and work that goes into making each type of evidence speak—or silencing it.

Notes

1. For an analysis of the tradition of Black witnessing that the complex dynamics of the circulation of these videos, see Allissa Richardson, *Bearing Witness While Black: African Americans, Smartphones and the New Protest #Journalism* (Oxford: Oxford University Press, 2020).

2. Mark Berman, John Sullivan, Julie Tate, and Jennifer Jenkins, "Protests Spread over Police Killings. Police Promised Reforms. Every Year, They Still Shoot and Kill Nearly

1,000 People," *Washington Post,* June 8, 2020, https://www.washingtonpost.com/investigations/protests-spread-over-police-shootings-police-promised-reforms-every-year-they-still-shoot-nearly-1000-people/2020/06/08/5c204f0c-a67c-11ea-b473-0-4905b1af82b_story.html. As I do the final edits on this chapter in 2020, people are protesting the police killing of George Floyd, caught on eyewitness video across the nation (and beyond). These protests—and demands for systemic change to policing—are a manifestation of the sort of political anger and energy that many have argued these videos ought to mobilize. It is as of yet unclear whether this video will have more evidentiary force in court than have many of its predecessors.

3. See, e.g., Peter Hermann, "Police Officers with Body Cameras Are as Likely to Use Force as Those Who Don't Have Them," *Washington Post,* October 20, 2017, https://www.washingtonpost.com/local/public-safety/police-body-camera-study-finds-complaints-against-officers-did-not-drop/2017/10/20/4ff35838-b42f-11e7-9e58-e6288544af98_story.html?utm_term=.270d20ab28da; Amanda Ripley, "A Big Test of Police Body Cameras Proves Ineffective," *New York Times,* October 20, 2017, https://www.nytimes.com/2017/10/20/upshot/a-big-test-of-police-body-cameras-defies-expectations.html; Nell Greenfieldboyce, "Body Cam Study Finds No Effect on Police Use of Force or Citizen Complaints," National Public Radio. October 20, 2017, https://www.npr.org/sections/thetwo-way/2017/10/20/558832090/body-cam-study-shows-no-effect-on-police-use-of-force-or-citizen-complaints.

4. Joel Snyder, "Res Ipsa Loquitor," in *Things That Talk: Object Lessons in Art and Science,* ed. Lorraine Daston (Cambridge, MA: MIT Press, 2004): 195–221; Allan Sekula, "On the Invention of Photographic Meaning," in *Thinking Photography,* ed. Victor Burgin (London: Macmillan, 1982), 84–109; Lorraine Daston and Peter Galison, "The Image of Objectivity," *Representations* 40 (1992): 81–128.

5. Sekula, "On the Invention of Photographic Memory."

6. Henry Fox Talbot, *Pencil of Nature* (1844–1846), quoted in Sekula, "On the Invention of Photographic Memory."

7. Objectivity; in this sense, photographs were positioned as, in Roland Barthes famous phrase, "messages without a code"—or better yet, for the natural sciences, messages inscribed in the code of nature herself (rather than in the codes of human culture and language).

8. Snyder, "Res Ipsa Loquitor."

9. The many complaints about the poor match between photographs and their objects—from those sitting for photographic portraits to police using photographs to identify subjects to scientists like geologists who found photographs useless for determining topography—in the late nineteenth century testify to the failures of photography to copy, or the gap between photographic representation and accurate representation at the time. Snyder, "Res Ipsa Loquitor," 213.

10. Daston and Galison, "The Image of Objectivity," 117.

11. Snyder, "Res Ipsa Loquitor"; Daston and Galison, "The Image of Objectivity."

12. Allan Megill, "Introduction," in *Rethinking Objectivity* (Durham, NC: Duke University Press, 1994), 1–20.

13. As Evelyn Fox Keller argues, the metaphor of aesthetic perspective is an apt one for capturing the paradoxes of objectivity in science. The technique of perspective drawing establishes realism by mimicking human perspective, in which objects in the distance are smaller than in the foreground, literally drawing from a particular viewpoint, though one that disappears in human appreciation for the realism of the scene. Thus, a view from somewhere is treated as the view from nowhere, resulting in the paradox of subjectivity: "in the vantage point of a particular somewhere at least the tacit promise of a view from nowhere." Evelyn Fox Keller, "The Paradox of

Scientific Subjectivity," in *Science and the Quest for Reality*, ed. Alfred Tauber (New York: New York University Press, 1997), 183.

14. Jonathan Sterne, *The Audible Past: Cultural Origins of Sound Reproduction* (Durham, NC: Duke University Press, 2003); Lisa Gitelman, *Scripts, Grooves and Writing Machines: Representing Technology in the Edison Era* (Stanford, CA: Stanford University Press, 1999).

15. Brian Winston, *Technologies of Seeing: Photography, Cinematography, and Television* (London: British Film Institute, 1996).

16. Science Friday News Roundup, National Public Radio, https://www.sciencefriday.com/podcast/hr1-news-roundup-marijuana-research-twitter-bots-chili-peppers/.

17. The shift in social sciences away from interpretation of social actions to measurement of observable behaviors is part of this move. Mark Smith, *Social Science in the Crucible* (Durham, NC: Duke University Press, 1994); Michael Schudson, *Discovering the News* (New York: Basic Books, 1981); John Durham Peters, *Courting the Abyss: Free Speech and the Liberal Tradition* (Chicago: University of Chicago Press, 2005).

18. Journalism is an instructive example of this confusion: there is no reason to presume that providing just two sides of a story will accurately, much less truthfully convey and event or issue. As the coverage of topics like climate change have showcased, this tactic can obscure as much as it illuminates.

19. Keller argues that in seventeenth-century science and philosophy, the view from nowhere aligned with God or the cogito. She suggests that in late twentieth-century discourse, an idea of "the system" (an idea anchored by numbers, statistics, and, increasingly, machines) has replaced God or the cogito as the vanishing point of the view from nowhere. Keller, "The Paradox of Scientific Subjectivity."

20. For more on the ways that the tradition of witnessing is embedded in embodied experience, and on the ways that truth in legal testimony (or witnessing) is tied to physical penalties, see John Durham Peters, "Witnessing," *Media Culture & Society* 23 (2001): 716. Peters goes on to suggest that the real scandal of post-structural critiques of representation (as always already interested) as not being the inability to neutrally document, but rather the suggestion that the way we arbitrate truth and reality ultimately comes down to bodily pain—a set of tests that the Enlightenment was supposed to have obviated.

21. Jennifer Mnookin, "The Image of Truth: Photographic Evidence and the Power of Analogy," *Yale Journal of the Law and Humanities* 10 (1991): 1–74; Jessica Silbey, "Cross-Examining Film," *University of Maryland Law Journal of Race, Religion, Gender and Class* 8 (2009): 101–31.

22. Commentators of the day decried the way that scientific expertise was aligned on each side to provide very different testimony. "The Howland Will Case," *American Law Review* 4 (1870): 625–54.

23. "The Howland Will Case." See also Snyder, "Res Ipsa Loquitor"; Thomas Thurston, "Hearsay of the Sun: Photography, Identity, and the Law of Evidence in 19th-Century American Courts," Roy Rosenzweig Center for History and New Media; http://chnm.gmu.edu/aq/photos/index.htm.

24. Snyder, "Res Ipsa Loquitor," 215.

25. "The Howland Will Case."

26. Mnookin, "The Image of Truth," 18.

27. Snyder, "Res Ipsa Loquitor"; Thurston, "Hearsay of the Sun." This legal discussion mirrors the emergent discussion of language as an arbitrary system of symbols in linguistics (the new science of semiotics).

28. "The Legal Relations of Photographs," *American Law Register* 17 (Jan. 1869): 4, 6. Cited in Thurston, "Hearsay of the Sun."

29. Mnookin, "The Image of Truth." Of course, this example demonstrates a way of thinking about photographic objectivity that goes far beyond the scientific discourse of mechanical objectivity described by Daston and Galison.
30. Mnookin, "The Image of Truth."
31. Mnookin, 65.
32. Many eyewitness videos are produced and circulated by Black bystanders to publicize police violence and killings; among the work these videos do is to provide believable testimony to distant White audiences of police brutality and the way Black men and women (but more often men) are policed differently from Whites. This is a dismal rhetoric: the need to turn to cameras as mechanical witnesses over their own testimony in order to be heard in White-dominated publics and media.
33. While some such videos are intended simply to document that an event occurred—say, the use of child soldiers or that an army was present at a particular time or place—the training on how to produce these videos by human rights groups emphasizes, as I will show, conveyance of perspective and identification even in these cases.
34. Some triggering systems are physical, electronic sensors built into the road. Others rely on continuous video feed monitored by computer programs that recognize movement of objects in the intersection during a red light, signaling the cameras to take a picture. "How Stuff Works," https://auto.howstuffworks.com/car-driving-safety/safety-regulatory-devices/red-light-camera1.htm.
35. The assignment of responsibility for the violation varies by states. Some takepictures that show the driver and ticket the individual pictured in the driver seat rather than the owner of the car.
36. The high cost of traffic violations in the Los Angeles area may be one reason why there appear to be more examples of legal challenges to red-light tickets in California than in other states. For example, even in Chicago, where a large number of red-light cameras were recently found to be malfunctioning, drivers seeking redress have followed political routes more than litigation.
37. *People v. Goldsmith*, 138 Cal Rptr. 3d, 305 (2012),: 310. Nonetheless, an Inglewood police officer in charge of the cameras (which were made and maintained by an Australian company, Redflex) was brought in to testify as to the validity of the images: that they did indeed represent the intersection in question and that the cameras and traffic lights were operating correctly.
38. *People v. Martinez*, 22 Cal. 4th 106 (2000): 117. The California Supreme Court decision, responding to an argument in the appeal about the representational status of the red-light photo (that there was a gap between what was shown in the photograph and the traffic infraction), argued that the inference could not be irrational, because it was the product of a computer.
39. This faith in computers was not only as devices of inscription (recording a faithful image of an event), but also as devices of storage and processing.
40. See *Search King v. Google* (2003); for more on the ways that automated speech fits within current First Amendment jurisprudence, see Toni Marie Massaro, Helen L. Norton, and Margot E. Kaminski, "SIRI-OUSLY 2.0: What Artificial Intelligence Reveals about the First Amendment," *Minnesota Law Review* 101 (June 2017): 2481–2525.
41. Authentication in these examples is purely technical: were the cameras operating properly, were the lights timed within the legally prescribed parameters, and were the storage systems working and maintained properly?
42. Kelly Gates, "Professionalizing Police Media Work: Forensic Video and the Forensic Sensibility," in *Image, Ethics, Technology*, ed. Sharrona Pearl (New York: Routledge, 2016), 41–57. I use the term computational objectivity in a way that overlaps with but is not identical to Gates's use here.

43. For example, the origins of regression analysis in eugenics. Wendy Chun, *Programmed Visions: Software and Memory* (Cambridge, MA: MIT Press, 2011); Jacob Gaboury, "Procedure Crystallized: Computation, Historicity, Theory," presentation at the Society for Cinema and Media Studies Conference, Toronto, March 16, 2018. For an overview of the types of bias in computer systems, see Batya Friedman and Helen Nissenbaum, "Bias in Computer Systems" *ACM Transactions on Information Systems* 14, no. 3 (1996): 330–47.
44. See, for example, Frank Pasquale, *The Black Box Society: The Secret Algorithms That Control Money and Information* (Cambridge, MA: Harvard University Press, 2015); Virginia Eubanks, *Automating Inequality: How High-Tech Tools Profile, Police, and Punish the Poor* (New York: St. Martin's Press, 2017).
45. Gates, "Professionalizing Police Media Work," 51.
46. For example, an Oregon man successfully challenged the math behind red-light camera citations in the state, and locals have been fighting corruption Chicago's red-light camera ticketing system for years. See Aarian Marshall, "Red Light Cameras May Issue Some Tickets Using the Wrong Formula," *Wired,* May 1, 2017; John Byrne, "City Reaches $ 38.75 Million Settlement in Red Light Ticket Lawsuit," *Chicago Tribune,* July 20, 2017.
47. For an analysis of how the video was made to say many things in court, see Frank Tomasulo, "I'll See It When I Believe It: Rodney King and the Prison House of Video," in *The Persistence of History,* ed. Vivian Sobchack (New York: Routledge, 1996). On the difference between White and Black reception of the video, see Elizabeth Alexander, "Can You be Black and LOOK at This? Reading the Rodney King Video(s)," *Public Culture* 7 (1994): 77–94.
48. It is possible that without the video, the police might have claimed not to have beaten him; however, given his injuries, this would have been hard to argue. The video thus did some work documenting that the beating occurred, but the locus of guilt or innocence moved to whether the beating was a legitimate or illegitimate action.
49. Kevin Bufford, "The Split on the Proper Review for Police Video Evidence," *American Journal of Trial Advocacy* 39, no. 2 (2015): 447–53; Marc Jonathan Blitz, Issue Brief, "Police Body-Worn Cameras: Evidentiary Benefits and Privacy Threats," American Constitution Society for Law and Policy, May 2015, https://www.acslaw.org/wp-content/uploads/2018/04/Blitz_-_On-Body_Cameras_-_Issue_Brief.pdf.
50. Bufford, "The Split on the Proper Review for Police Video Evidence."
51. Alexandra Mateescu, Alex Rosenblat, and Danah Boyd, "Police Body-Worn Cameras" (working paper), Data & Society Research Institute, February 2015, https://datasociety.net/pubs/dcr/PoliceBodyWornCameras.pdf. Howard Wasserman lines up preliminary evidence that lower courts have been treating video evidence differently (allowing more inferences from the video) since *Scott v. Harris.* Howard M. Wasserman, "Orwell's Vision: Video and the Future of Civil Rights Enforcement," *Maryland Law Review* 68, no. 3 (2009): 600–661.
52. The majority found that "no reasonable juror" could find differently than they did. They did not feel the need to respond to Justice Stevens's dissent, in which he pointed out the possibilities of interpretation in the video. Rather, Justice Scalia uploaded the video (along with his concurrence, posted to the Supreme Court website), apparently as a statement of incontrovertible fact. See Silbey, "Cross-Examining Film."
53. Others have commented on the cinematic aspects of the video and the way that the location of the dash cam invites the viewer to take on not only the visual perspective, but the subjectivity of the police engaging in the chase. See Neal Feigenson and Christina Spiesel, *Law on Display: The Digital Transformation of Legal Persuasion* (New

York: New York University Press, 2011); this commentary on the relation of aesthetics to ideology echoes much work in film scholarship.

54. *Scott v. Harris*, 500 U.S. 372 (2007).

55. Dan Kahan, David Hoffman, and Donald Braman, "Whose Eyes Are You Going to Believe? *Scott v. Harris* and the Perils of Cognitive Illiberalism," *Harvard Law Review* 122, no. 3 (2009): 838–906.

56. The campaigns I look at differ in their scope and focus; some are aimed at international courts, others at U.S. state or federal courts.

57. Forrest Stuart, "Constructing Police Abuse after Rodney King: How Skid Row Residents and the Los Angeles Police Department Contest Video Evidence," *Law and Social Inquiry* 36, no. 2 (2011): 327–53.

58. Since the advent of light-weight filming equipment for use in television news and documentary in the 1960s, such long shots have become a register of realism and an index of the filmmaker's presence at the scene, with any shaking or movement of the camera understood not as part of a cinematic apparatus, but rather as a reminder of the body behind the camera, at the scene. (The hand-held camera effect, and its reference to the indexicality of the moving image, has of course become part of the apparatus of cinema, used as an effect to evoke a particular realism, as in horror films like *Cloverfield* [2008].)

59. Stuart, "Constructing Police Abuse After Rodney King." As Stuart documents, this strategy arose expressly in response to the use of video like that of the beating of Rodney King to bolster police narratives.

60. The organization promotes successes it has had in providing evidence to the International Criminal Court that have assisted in the prosecution of several warlords in Africa. "Our Work," WITNESS website: https://witness.org/our-work/.

61. On the work of WITNESS to create experts and an infrastructure for making eyewitness videos legible as visual evidence, see Sandra Ristova, "Human Rights Collectives as Visual Experts: The Case of Syrian Archive," *Visual Communication* 18, no. 3 (2019): 331–51.

62. "Basic Practices: Capturing, Storing and Sharing Video Evidence" and "Filming Secure Scenes," Video as Evidence Field Guide, WITNESS, https://vae.witness.org/video-as-evidence-field-guide/.

63. WITNESS, "Filming Secure Sites" Field Guide (pdf): www.mediafire.com/view/dgv84df8ldwz1ba/VaE_FilmingSecureScences_v1_0.pdf; "Video as Evidence: Documenting Standing Rock" (pdf): https://library.witness.org/product/video-as-evidence-documenting-standing-rock/; "How to Film Protests: Filming Tips" (video): https://www.youtube.com/watch?v=e1f2Prk1cSQ.

64. One of the powerful ways that eyewitness videos may operate is in the way the publicity they generate may (slowly) train legal interpreters (from judges to jury members) to identify with eyewitnesses and victims in these videos rather than with the police officers.

65. Keller, "Paradox of Scientific Objectivity," 183.

CHAPTER 2

Eye-Tracking Techniques and Strategies of the Flesh in *The Brother from Another Planet*

Notes toward a Visual Literacy of Video-Recorded Lethal Police-Civilian Encounters

KELLI MOORE

Introduction

For decades, exonerations of police officers accused of fatal overuse of force against civilians of color have persisted, inciting insurrectionary and fugitive forms of public assembly. In particular, reactions to the killings of unarmed Black civilians raise questions about the relationship between US police forces and the history of slavery. Video recordings of unarmed Black people dying constitute a repeating body in which slavery lives its afterlife across the gamut of media forms.[1] In this macabre social media environment, many have struggled with the moment in a text, tweet, or other broadcasting platform when it comes to listing the names of the dead in a way that publicly redresses the importance of the unique individual lost to state violence while also acknowledging the larger structural significance of anti-Black violence as world-making. Along these lines, it has been crucial to work on and work through the gendered nature of state violence toward people of color, as seen in the #SayHerName social media campaign that draws attention to the structural frequency of Black, Brown, Indigenous, and trans women killed, raped, and beaten by police forces.

Critical race and law scholars have long known that anti-blackness is the logic of desire of the police and much of the Anglo-American legal tradition. Saidiya Hartman's description of Black cultural expression identifies a variety of slave laws that position the freedom of Black movement and vocalization at the whims of the master's whip.[2] In the post-civil rights era, Kimberlé Crenshaw's attention to labor and employment case law demonstrates that law has difficulty in recognizing the legal subject as both Black and female, raced and gendered.[3] Crenshaw's intersectional analysis is a mode of bringing attention to law's blind spots. Her argument accords with Eduardo Bonilla-Silva's observations of the hegemonic discourse of the "colorblind society" that functions as the alibi for structural and institutional racism where anti-blackness is simply projected as a rare negative outcome involving a "few bad eggs."[4] More recently, Sora Han gets at the root of the gaslighting on anti-blackness by tracing the route of psychoanalytic desire and the role of the unconscious to understand the ways law and its subjects are read through colorblind fantasy.[5] Through a reading of the law's letters—what law says—Han shows how anti-blackness is the logic of law's desire.

Despite what critical race and legal scholars know about how anti-Black racism lives in the letter of the law, there are calls for more interdisciplinary conversations about how anti-Black *legal* thought lives in visual culture. Though we are increasingly attuned to the reality of anti-blackness as the dominating logic of law's desire, a number of theorists also point to the new media context in which the law is anchored in visual and digital tools and simulations. This is especially germane to video recordings that capture anti-Black racism "in the act." A plethora of bystander videos and popular art forms circulate about Black Lives Matter (BLM) protests that foster debate about the public consumption of Black pain and suffering. Scholars have not merely pointed out the dearth of conversations about how images and video of black people detained and murdered by police should be read; they have also situated the absence of an interpretive framework within sophisticated mechanisms of advertising revenue streams, news ratings, and other aspects of capitalist production.[6] Those who capture such video content are also acknowledged as racialized media laborers and producers.[7] For example, many have begun to question the meme-ification of Breonna Taylor's murder by Louisville police on March 13, 2020. Quotidian mentions of Taylor's name on social media seem to trivialize her death and the larger structure of anti-blackness; however, as journalism scholar Allissa Richardson finds, meme-ification "may seem like it's a joke or using satire, but it's really just a method to trick the algorithm to talk about [Taylor] again."[8] As social media companies engineer private and state surveillance under the cover of free and open communicative interaction, Taylor's case critically indicts the ways anti-blackness also resides in computational

algorithms that limit the appearance of social media posts about the controversy over Black deaths at the hands of police.[9] In addition to the problem
of consuming Black and Indigenous suffering is a growing debate about
the distinction between law and culture, exemplified by the simulation of
legal evidence by public news outlets. In another example, the *New York
Times* reconstructed a timeline of events leading to the murders of Ahmaud
Arbery and George Floyd using multiple surveillance and bystander videos.
These videos circulate what appears to be visual evidence in the public
sphere in ways that mimic the representational techniques utilized by law
officials to display visual evidence during criminal trials. Together, the visual
culture of law and the history of anti-blackness inform the digital media
context in which BLM protests of the murders of Black civilians by police
disclose the relationship between the law and the visible. Yet a question of
intersubjectivity remains concerning how the individual's apprehension of
the field of vision is transformed into the certainty of witnessing "racism
in action" that can direct legal fact finding and decision making.

Visual evidence collected from bystanders who record fatal interactions
between police and Black civilians have failed to offer hard proof of police
overuse of force that would corroborate the anti-Black structure of US
policing. Further, incorporating cameras into the police uniform and equipment (e.g., police dashcams), which occurred in several police departments
around the United States prior to the previously mentioned high-profile
killings, has not emerged as the solution many hoped. Body-worn cameras
are plagued with the problems that often befall institutions and organizations implementing new technology among workers and other stakeholders:
lack of technical support for police required to use the cameras, lack of
oversight concerning the implementation of the cameras, inconsistent
camera use, and an absence of clear and swift consequences for officers
who fail to record interactions with civilians where force is deployed.[10] On
top of these dilemmas, police departments that implemented body-worn
cameras have written few if any policies that would regulate their use.

In this context of the failure of video evidence to confirm instances of
police brutality and the anti-Black structure of police work, this chapter
examines a machine-learning technique used in criminology research,
known as eye-tracking, for the ways it discloses the future relationship
between law and the visible. Eye-tracking is a form of machine learning used
to surveil how much time spectators attend to information presented in the
field of vision. In the social sciences, laboratory eye-tracking has become
a significant tool for drawing inferences about how spectators interpret

video-recorded exchanges between police and Black civilians. Given law's discovery of new protocols for its own surveillance through video footage, whether appropriated from video-recording bystanders, closed-circuit television, or police dash cameras, perceptual bias experiments are increasingly significant for what they illustrate about potential jurors and courtroom audience members. Perceptual bias research of video recordings that capture anti-Black violence "in the act" heavily relies upon the use of eye-tracking techniques by social scientists. This method has earlier roots in visual culture studies. When anthropologist Charles Goodwin deconstructed the rhetorical deployment of visual evidence in the Rodney King police brutality trial, his analysis of the defense team's use of slides led to the influential finding that members of institutional communities and groups are gathered, trained, and organized into their own ways of seeing.[11] The ability to offer one's "professional vision," i.e., expertise in a particular mode of seeing, is endowed by one's membership to a community of practice. In the recent police overuse-of-force cases in Minnesota, Kentucky, Georgia, Missouri, Oklahoma, Wisconsin, Ohio, South Carolina, New York, and beyond, how do juries see and how are they led into sight?

Over twenty years ago, legal scholar Reva Siegel asked, "What or where is the legal phrase 'in the eyes of the law?'"[12] The tradition of this scopic metaphor illustrates how sight is ascribed to law; it signals law's all-seeing capacity and opens analytical pathways to just how law maintains its commitment to the primacy of its own vision. Not only do the law's eyes embody authority, they do so by adding a corporeal dimension to law's essence, to the status of law as ontology. More important, the authority of the eyes-of-law metaphor affirms disinterested, objective vision, a pure form of sight invulnerable to imperfect vision wrought by bias, predilection, prejudice, and inclination— precisely the issues at play when video-recorded interactions between police and civilians lead to accusations of overuse of force. By considering the inferences made through eye-tracking techniques, this chapter suggests that law may increasingly understand that anti-Black racism is a form of embodied knowledge, conditioning bodily functions of seeing, looking, and witnessing. Further, I comment on the conditions and populations for whom the pursuit of fact-based justice will be reshaped and respatialized when anti-blackness is understood as a form of embodied knowledge.

After a brief history of the eye-tracking device, the chapter analyzes how the technique objectifies the work of human attention. I show how attention, where the eyes roam and rest during the act of looking, is rendered an object of legal analysis by eye-tracking techniques. Scientific inferences about attention open a new set of inquiries that lie at the intersection between law's use of visual evidence and the architectural environment in which evidence is displayed. Scientific papers are briefly reviewed for the ways eye-tracking techniques objectify attention by breaking down eye movements in terms "optical fixation." In one of these papers, optical fixation evinces two experimental outcomes researchers call "Attention Divides" and "Black Sheep Effect," which are also described. Readers of law and literature may struggle with the technical emphasis of this section. Let me preempt any potential difficulty by briefly addressing the stakes of deconstructing eye-tracking technology. As scholars have shown, the algorithm's computational language operates as both law and literature and is constitutive of the eye-tracking technique.[13] I argue through a cinematic example that the technical affordances of eye-tracking are descendants of *legal emblemata*, iconic representations of law, notably examined by Peter Goodrich.[14] A detour to John Sayles's 1984 film *The Brother from Another Planet* helps situate contemporary eye-tracking techniques within the metaphorical and figurative slave history of the "eye of law."[15] Finally, I also use the film to demonstrate how scientific inferences enabled through eye-tracking suggest the reality of how anti-blackness is lived in and as law. I conclude that positioning the eye-tracking technique within the genealogy of legal vision suggests the future relationship of law and the visible will be computationally mediated but should be guided by dark sousveillance, a practice of counter-vision found in Black freedom struggles.[16]

Legal Spectatorship

The stakes of the following arguments concern members of the courtroom audience. Although professional vision can tell us about how prosecutors and defense attorneys displayed and discussed video evidence, it does not tell us how courtroom audiences perceive evidence. The conditions through which we see fatal police-civilian interaction is part of a special form of looking I call legal spectatorship. Legal spectatorship "delineates looking practices for courtroom testimony called into being by the inter-action of the First and Sixth Amendments of the US Constitution."[17] This concept encompasses the relationship between law and visual literacy beyond realms traditionally associated with law's courtroom environment.[18] "Constitutionally protected through interlocking terms of the Sixth and First Amendments," legal spectators are "instantiated as a central and permanent crowd of witnesses whose spectatorship, conducted from this optimal classic viewing position in the space of the courtroom, both materially and symbolically fulfills the individual's democratic right to a public trial."[19] The spatial situation of the audience also "affirms the public's combined rights of free speech, free press and free assembly."[20] Legal spectatorship makes explicit the obligatory nature of courtroom looking practices where looking is a public duty through which citizenship is ritually performed. The concept also raises methodological questions and concerns about how the public discovers legal facts from visual evidence in the absence of discussion of literacy about such material. To what kind of visual literacy do we assume the public has access? Is there an implicit value to the silence on visual literacy on the part of constitutional amendments that codify free public assembly and a trial of one's peers? Borrowing a question from legal theorist Marianne Constable, is this silence just?[21] And could such a doctrinal silence be sustained given the saliency of skin color revealed in punishment decisions? Many are equally invested in the freedom to assemble publicly and ensure that US citizens stand trial in public rather than under cover of secrecy. These are crucial forms of spectatorship codified by law.

Yet the frequency of deadlocked juries, mistrials, and acquittals in police brutality cases suggests no standard exists that would guide how law professionals circulate such materials or how court audiences perceive the same. This problem is not only manifested in the absence of protocols for police body and dashboard cameras; it is also manifest in the cinematic representation of automatic reporting devices (also known as machine learning), law, vision, and their structural connection to anti-Black racism and the history of slavery. Legal spectatorship, then, is not only a de jure looking practice. In this chapter I focus legal spectatorship on de facto looking practices that have stakes for criminal jurisprudence. Posing these questions affirms the general anxieties that automation and predictive technologies pose to law and culture while also acknowledging the literal and metaphorical relationship the "eye of law" and "blind justice" constructs between law and the visible.

A Brief History of Eye-Tracking Technology

Eye-tracking first emerged in the nineteenth century as a way to study cognitive transformations during the act of reading text. They began as stationary and intrusive apparatuses that combined a contact lens with a cutout for the pupil that was connected to an aluminum pointer that moved in time with the movements of the eye. Its technical development

occurred over a number of technologies from electro-oculography, scleral
contact lenses and search coils, photo- and video-ocularity, and reflective
devices.[22] Eye-trackers evolved into lighter, portable, and less intrusive digital
mechanisms that reflect beams of light off the eye, which are then recorded
on film. Today, tracking what the eye does while reading is computationally
mediated and mapped. Rather than a given text that simply offers up its
meaning, tracking techniques reveal that the human eye performs a lot of
work to construct meaning through erratic, jerky movements filled with
pauses, false starts, and back traces. In addition to rendering visible the
eye's perceptual processes through computational constructs, the technique
also establishes gaze preferences.

Eye-tracking techniques have evolved into contemporary applications in
employee training, market research, and product development. Scientific
research on eye-tracking techniques describes their importance to neuromarketing and technology interface design.[23] Outside the realm of consumer
product development, the technique is a primary example of technology of
transparency, a Foucauldian term applied in feminist science technology and
society studies (STS) to those techniques that play a role in the construction
of subjectivity. Likewise, eye-tracker devices may also be characterized as
a transparent technology and critiqued for its deployment of an objective scientific gaze that cannot produce knowledge of the body without its
disassembly.[24] Feminist STS demonstrates the gendered effects of such
technologies, for example, in the polygraph and its ability to "show" the
presence of lying. In the case of the polygraph or lie detector, feminine
conscience is made transparent by technically rendering, or objectifying,
attention. The origins of the polygraph are part of a national obsession with
understanding the female mind and its so-called tendency to lie.[25] Like
the polygraph, eye-tracking devices are implicated as techniques of truth
that recapitulate the Cartesian mind-body division. The devices are part of
the techno-cultural history of making attention and conscience visible by
disassembling the body, isolating it into part-objects.

These tools render attention visible through idea of "optical fixation."
Generally, the concept of fixation occurs in psychoanalysis, denoting the
emergence of abnormal sexual traits, for example, voyeurism in the work
of Freud. A look at perceptual bias research findings suggests how optical
fixation and the psychoanalytic discourse of fixation coalesce. Polynomial
(i.e., algorithmic) expressions calculate more than the number of optical
fixations; they also calculate duration. Studies using the method divide
fixations such that their number illustrates the engagement of attention
while the duration of fixation indicates a difficulty disengaging attention.

Optical Fixation: Calibration and Calculation

In university criminology and psychology laboratories, eye-trackers are stationary objects connected to computer screen displays. Most eye-trackers employ a video-based system that measures the movement of the cornea and pupil. This process works through reflection where infrared light is reflected through a mirror into the subject's eye. A reflection off the cornea and retina is what will emerge as the optical fixation. The location of the participant's eye is calculated from the corneal glint and the retinal reflection.[26] In combination with the algorithm, the eye-tracker is an automatic recording device that reports human vital phenomena. Optical fixations may be further distinguished as observations that "occur each time a specific cluster is entered and exited."[27]

In social science experiments, the gaze is staged as a vital activity whose transformations may be recorded and calculated algorithmically. Polynomial expressions encode, approximate, define, and delineate functions taken from the gaze. They register or make apparent eye position as an utterance, via instantiations of mathematical law. Two approaches define eye-tracking algorithms: feature-based and model-based. The feature-based approach operates by an individual's unique eye features. The model-based approach operates by designating the best-fitting image of a model eye.[28] Investments in the technology are made via transfers at the level of scholarly publication. New journals are devoted to the evaluation of eye-tracking technology, particularly around how to calibrate the apparatus and the relationship between statistically significant results relative to the cost of eye-monitoring technology. Scientific journals are also evaluating the performance of particular algorithms for specific calculations. The Starburst algorithm has emerged as a particularly robust one in studies of perceptual bias, for example.

Calculating the optical fixation is performed at a critical interdisciplinary nexus between the social sciences, humanities, and mathematics/computing. The eye-tracing apparatus is a black box—no printout emerges from the connected devices. The interpretation of the *mise en scène* of optical fixation—the computational construct in which fixation phenomena is produced—is illustrated in published scientific papers. Though the scenes in which optical fixation is calibrated are reported, its visual representation occludes the fact that in experiments of perceptual bias optical fixation must be surreptitiously collected/calculated. Algorithmic calibration captures the gaze patterns in the perceptual bias laboratory with the subject digitally tethered to the device. Optical fixation is a phenomenon of a networked (calibrated) polynomial's mathematical transformations. These transformations are enabled by the creation of a unit of analysis, the "eye pixel." The compression between those terms "eye" and "pixel" illustrates a parasitic link between human and computer.

Visual Evidence of Anti-Blackness in Scientific Journals

Research relying on optical fixation suggests evidence for the "attention divides" hypothesis. "Attention divides" refers to the phenomena in which a subject "focusing on a common target will exaggerate bias in punishment decisions among individuals who vary in identification."[29] The authors further flesh out the hypothesis writing:

> People direct visual attention towards people that they consider to be threats. . . .
> Likewise, White Participants directed attention to Black faces over White faces,
> but this attentional bias was eliminated when the faces displayed averted gaze,
> which reduced the threatening nature of the faces. In our studies, participants
> may have considered police and outgroup members threatening, which may
> have led some to orient attention to them. Moreover, our videos depicted
> physical altercations, which evoke feelings of realistic threat. Expectations
> for violence from outgroup members during altercations and motivations to
> be vigilant for aggression from outgroup members may lead people to direct
> attention to the source of threat, shape understanding or case facts, and ulti-
> mately, lead to harsher punishment decisions.[30]

In contrast to "attention divides," researchers also find the "Black Sheep Effect" produced in the laboratory, in which out-group identifiers punish hypothetical out-group targets more harshly than weak out-group identi-fiers.[31] Crucial to these findings is how study participants intensify pun-ishment of in-group members they perceive to be deviant. Experimental protocols stimulate participants not only through typed narratives of a hypothetical disciplinary situation. The findings are driven by the inclu-sion of video simulations of police-civilian interactions into the research protocol. Eye-tracking techniques are then applied to study participants' gaze patterns. Optical fixations are the data that disclose "attention divides" and the "black sheep effect" as criminological hypotheses, both part of an anti-Black ethos of punishment. Perceptual bias studies suggest that the experience of seeing images replayed may increase punishment decisions among spectators. Eye-tracking devices are technologies of vision that track more than the mere fact of reading difference that becomes racialized. The scientific object—optical fixation—further renders the anti-Black character of racial difference, the significance of scene repetition, and its imbrication in the spectator's future punishing behavior.

Suspend the following metaphysical questions about the materiality of optical fixation in order to consider recent jurisprudential hypotheses they underwrite, for the findings are rhetorically compelling despite an ambivalent orientation to the fixation phenomena upon which such research depends: What is the materiality/visuality of the optical fixation as a vital report? What evidence of thought and perception does the technique confirm? How sophisticated does the apparatus need to be to achieve accurate calculations? Whatever the ambivalence about eye-tracking techniques and the calculation of incalculable, the technique makes a visible scientific object out of human attention.[32] Eye-tracking brings the optical fixation and many provocative psycho-physical phenomena into the genealogy of legal vision. They do so through the pervasive language of computation.

I want to take up further the position this form of machine learning occupies in the history of techniques of transparency and truth. Technologies of vision have long played a role in constructing knowledge of femininity and gender difference. Likewise, knowledge of racial difference has always been a part of the visual discourse of truth and is now enabled by machine learning as a matter of routine. The dominant idiom of the algorithm, manifest in the optical fixation and punishing decisions of study participants, demonstrates that the structure of the field of vision and, by extension, of visual competency is always already anti-Black. On a material level, eye-tracking methodology highlights the prevalence of video footage as the dominant matter in criminal jurisprudence. There is a blending of the concepts of film and skin ascendant here, where bystander and police dashboard videos are projected by the public at large and by the state as a strategy of the flesh. The optical fixation discloses the eye's attention (where attention stands in for the political philosophical concept of recognition) to the flesh in a visual field, whose movements are located within the film itself and the spatialized projections of courtroom and personal screens. Previously, scholars interpreted how the defense in the Rodney King police brutality trial broke down the bystander footage, broadcast around the world, still-by-still to construct King as the crazed Black figure controlling the interaction with endangered White police.[33] Eye-tracking methods contribute to a visual pedagogy in a jurisprudential moment in which digital bystander video recordings of lethal police-civilian encounters proliferate. The technique reorients attention away from how the defense displayed visual evidence in the King trial to focus on the spectator's preexisting anti-Black viewing apparatus.

Richard K. Sherwin has commented on the imbrication of law and culture, aesthetics and ethics, arguing that law's visual life exists within a larger cultural abyss of anxiety and doubt concerning the veracity of the image on one hand and the excess of meaning conveyed by images on the other:[34] "We need a new visual literacy to crack the aesthetic, cognitive, and cultural codes of law as image in the digital age."[35] Eye-tracking techniques expose colorblind racism as the aesthetic, cognitive, and cultural code of law by deploying the discourse of the algorithm to verify anti-blackness as constitutive of both the field of vision and the spectator's interpretation of its recorded activity. The technique suggests there is no such thing as a nonracist or race neutral spectator of law. Colorblind discourse, then, is the alibi for anti-Black practices of looking in law and culture. In this sense, the call for better visual competency of law must conceive of the eye of law and notions of blind justice anew.

In their volume *Genealogies of Legal Vision*, Peter Goodrich and Valérie Hayaert observe that far beyond the letter of law or legal text, law is made accessible through many architectural structures common to the courthouse: "columns and steps, bifold doors and ornamented foyers."[36] These openings serve as portals to law. Further they contend,

> The portal to the law is itself an emblem, an image of entry, an opening of the curtain onto the tableau and stage of the juridical. Thus, the figure of law that inaugurates the body of laws is that of a higher power, a super-terrestrial image, such as that of Leviathan for Hobbes or that of the divinity, depicted as a crescendo of light, handing the laws to Moses, to Solon, to Homer, or some further figure of mythological or Christian rule.[37]

Soon after, the authors draw attention to the face as the locus of law's encounter, its portal. They write:

> The face is an opening; it is the conduit of breath in Augustinian terms of *pneuma*, of breath and of all the other insufflations that are marked by the diverse facial orifices that allow for the passage from external to internal, from visible to invisible. There is importantly, a dual relay connected to the face in that it opens to and simultaneously or also lets in, imbibes and exhales. If the face is an image, a mobile site of identity and difference, the image is equally a face and in the case of law, it is explicitly *facies altera* [back face, or two faces], the mark of time and judgment.[38]

"I can't breathe," the common refrain improvised by protests against the anti-Black structure of law and policing, suggests the need for an alternate portal into law than the face or body that is more explicitly future-oriented than Goodrich and Hayaert's analysis. The civilian's inability to breathe due to the weight of the police officer is not merely physical but a historical and metaphorical manifestation of Western politics. Their conception of law's two faces (*facies altera*), rather than explicitly indicating the colorblind neutrality of the ancient Western juridical tableau, is in danger of eliding the explicitly anti-Black structure of law and the "higher power" that serves as guarantor. While I take the author's point, I want to pivot away from the Eurocentric image and face of law they draw and instead imagine the physical and metaphorical encounter with law as a looking practice comprising a strategy of the flesh. In the next section, I examine the film *The Brother from Another Planet* for how it marks the alterity of time and judgment in order to reimagine law's visual competency that attends to Black feminist inquiry about the flesh.

The Brother from Another Planet

While the use of eye-tracking techniques to study perceptual bias and the incitement to punish positions human attention and bystander video recordings as objects of critical legal pedagogy, disciplining the eye of law is frequently modeled in popular cinema. The ideal representation of legal knowledge is borne out in the history of the eye and is prominently featured in the science fiction genre. In this genre, what happens to people's eyes is suggestive of the future of legal pedagogy. Famous examples, namely *A Clockwork Orange* and *Bladerunner*, represent the future entanglement between law and the visible. Recall how the Ludovico eye technique and Voight-Kampf methods, respectively, used in these films represent the eye of law in terms of disciplinary apparatuses of the corporate state.[39]

The Brother from Another Planet stages a break from the eye of the law legal fiction by dispensing with the mechanical representation of the transparency apparatus to an embodied version, one where the eyes of the law reside in the Black alien male. *The Brother from Another Planet* is about an alien escaped from origins unknown, played by the Black American actor Joe Morton. The film explores two dualities: one between the inseparability of Black flesh from body and the other being how "alien" references both immigration status and extraterrestrial being. Alienated in this world, Black flesh signifies extraordinary commodity value. The Brother, for example, regenerates his lower leg and foot that was severed upon his entry to Earth, hitting upon the historical association between blackness and hyper-analgesic qualities and endless rejuvenation. Such qualities are those precisely at issue when the cries of Black civilians are ignored by police officers. These are remarkable abilities the protagonist sells to electronics stores for cash under the table. The Brother lives his life as a mute boarder in a Harlem apartment building, communicating through nonverbal gesture to neighbors, and is pursued by two White intergalactic immigration offices, Men in Black. The Brother is illegal, and his crime is being.

The Brother's feet present a strategy of the flesh in which the foot embodies his difference from the humans upon whose planet he has crashed landed, a few clicks from the Statue of Liberty. The Black community adopts him, presumably based on shared phenotype; however, the Brother's strange feet will walk an alternate path toward the making of law and punishment. The film introduces the foot as evidence of community membership into the hieroglyphics of legal emblems. Another example of the film's revelation of flesh is the disembodied immigrant and refugee voices the Brother can hear from the past, the many tongues inhabiting the halls of Ellis Island. The Statue of Liberty and Ellis Island suture a relation between freedom and community difference whose path the Brother will traverse in the film.

Reva Siegel argues that legal fiction, such as the eyes of law, "may itself be a figure of speech that naturalizes the rich variety of ways that the language of the law constructs the social world we inhabit."[40] While continuing to evade the Men in Black, the Brother conducts his own investigation into the corporate-sponsored drug traffic that is killing his neighbors in Harlem. One afternoon his surveillance of local drug dealers is interrupted by the presence of a police officer. The Brother dislodges his right eye and places it in a planter, angling it toward the dealer's building. Once removed from his body, the eye functions as a pulsating surveillance camera whose footage he can reinsert into his body to play. Later, when the police officer has gone, he recovers and replays the footage by returning the eye to its socket; the Brother becomes a video recorder, playing back the film collected by his detachable camera-eye. It is through corporeal disassembly and reassembly, taking apart a piece of the body and putting it back together, that the Brother locates a fugitive and insurgent "eye of law" within Black interiority. This is a position counter to how the state deploys "eyes of the law" through the evaluation of Black exteriority in ways that subject Black people as a class to legal exceptions that are frequently fatal and always stigmatizing and victimizing.

Staged within the film is a distinction between flesh and body made by cultural critic Hortense Spillers in her widely influential essay, "Mama's Baby, Papa's Maybe: An American Grammar Book."[41] Spillers's formulation implicates both colonial and slave laws that legitimized the systematic use of violence by the master class. The authority and legitimacy of the hieroglyphics of the flesh, the scars meted out by White masters upon Black slaves, were gradually transferred to the hands of police after the Civil War, stretching to contemporary everyday interactions between police and Black and Brown civilians.[42] *The Brother from Another Planet* portrays this afterlife of slave fugitivity. Peter Goodrich's erudite description of legal emblems is also concerned with the hieroglyphic. His analysis of Anglo-European legal desiderata decodes the visual hieroglyphs that construct the values, powers, righteousness, ambiguity, and the authority of Western governance.[43] Hovering eyes, hands, pierced hearts, and other symbols abound in emblem books that visualize how law shall govern. These media forms and their hieroglyphic messages were encoded for those individuals initiated into governance roles by status. The visual culture examined in his text ranges from the mid-1500s to the late 1600s, the opening century of the transatlantic slave trade.[44] While this fact goes unmentioned by Goodrich, his analysis nonetheless helps to lay bare the other side of the visual hieroglyphics Spillers draws upon to inform her theory of the violent process of Black enslavement. "Lawyers played a crucial role in the structuring of vision, in making power visible and so providing models, patterns, and in short accessible exempla, emblems of the licit modes of disposition and behavior" through these media forms.[45] Guided by Spillers's emphasis on the way enslavement makes flesh out of the person, the models' patterns of law were made visible in more than the legal emblem book. The anti-Black logic of the Anglo-European legal and cultural tradition is borne upon Black body rendered as mere flesh. Thus, the production and circulation of this cult classic film illustrates the hieroglyphics of the flesh Spillers named decades ago and the hieroglyphics of the emblem book Goodrich deconstructs more recently.[46]

Yet one of the interventions of Black feminists who keep fugitive struggles in mind has been to expand Spillers's notion of the violence against the body. In this way, anti-blackness is imposed through strategies of the flesh that take seriously the alterity subjects of anti-blackness, who often employ it to create knowledge about self and word, including law and visual culture.[47] Consider Simone Browne, quoting Steven Mann: "Before approximately 50 years ago—and going back millions of years—we have what we call the 'sousveillance era' because the only veillance was sousveillance which was given by the body-borne camera formed by the eye, and the body-borne recording device comprised of the mind and brain."[48] Dark sousveillance can be an organized and improvised strategy that "entails an active subversion of the power relations that surveillance entails."[49] Its fleshy strategy may assume many material and ephemeral forms. The recording by Darnella Frazier of the suffocation of George Floyd by a Minneapolis police officer is another example of sousveillance in a long line of bystander videos.

FIGURE 1. Rookie cop speaks to the Brother, who is rigid with discomfort. Courtesy of the author.

FIGURE 2. The Brother deposits his eye in the planter. Courtesy of the author.

FIGURE 3. The Brother's eye sousveiling the neighborhood. Courtesy of the author.

These flesh-film recordings function as citizen undersight in a visual field dominated by state and corporate oversight. *The Brother from Another Planet* and eye-tracking techniques converge at the bystander recording to imagine the kind of visual literacy legal spectators will develop to read them in and outside courtroom settings.

On the stoop of a brownstone, the Brother has a one-sided exchange with a White rookie police officer. The officer relays that his colleagues have warned him about the racialized perils of his beat: "How long you been up here, pal? You a native or what? My first day in this precinct. Partner upstairs been handing me all kinds of horror stories like, you know, they are going to cook me alive if I don't look out what I'm doing. Can't be all that bad I figure? Pretty damn nice to me, Harlem. You know people are people, right? You put on a uniform, it's not like you hand in your status as a human being. I mean we're here to protect and serve, right? Like it says on the LA cruisers on Adam 12. They never should have taken that show off the air. So, what do you think?" The Brother does not engage the officer's words; instead, he silently walks away.

As the Brother shuns the officer, he leaves his alien eye behind to record the community on his own. He models a strategy of the flesh in which the separation of the "eye of law" from its Black alien inhabitant discloses a distributed and doubled model of knowing justice from that of the police offer. The Brother's detached eye is a fleshy example of dark sousveillance. As cultural critic Tavia Nyong'o imparts about the film, the Brother's eyes are an "emblem of black counter-surveillance."[50] The Brother's eye is fixated on his community. Once returned to his body, his recordings become embodied knowledge that counter the forms of knowledge generated by eye-tracking techniques in the implicit bias experiments discussed earlier. *The Brother from Another Planet* prefigures the use of eye-tracking technology in a way that brings the fleshy strategy of the Brother's counter-sousveillance into conversation with the kind of nonrepresentational vision understood experimentally as optical fixation. The Brother has indeed plodded a different path with his odd feet. Between his detachable eye and optical fixation, the hieroglyphics associated with the "eye of law" found in legal emblemata are extended to celluloid, digital film, and other computational constructs.

Conclusion

What will be the doctrinal significance of the human attention system, rendered knowable through eye-tracking techniques? This chapter has performed the bridgework necessary to further link critical race and legal studies to visual culture and media. Digital bystander footage is a political and social form of visual evidence: political because it implicates the state and police in centuries of long and violent repression of Black freedom, and social because it performs solidarity among the bystander, a member of the viewing public, and the victims of police lethality through the cinematic apparatus. The vernacular media practices that bring digital bystander footage into being thus resist interpretations governed solely by traditional indicators of film genre or histories of photography. Detailing the use of eye-tracking techniques in experiments about implicit bias in police-civilian encounters illustrated the optical fixation's status as a media form and technique rather than a genre element in visual studies that concern law.

Although eye-tracking and optical fixation are presaged in the science fiction genre, I have suggested, through Black feminist theory of the flesh, the significance of film as a form of flesh deployed strategically to resist anti-Black logics that inform the eyes of law and blind justice tropes. *The Brother from Another Planet* depicts this flesh-bound strategy of dark sousveillance where community undersight challenges the authority and legitimacy claims of state and police oversight. Here, state and police oversight include police body and dashboard cameras offered as a reform measure to placate an increasingly angry public. I want to make clear that I am extremely dubious that implicit bias training or body and dashboard cams for police will be helpful. For these are reform measures, repairs that do not redress the logic of anti-blackness of Anglo-American legal and cultural tradition. In other words, the state's decision to record its own lethality is not constitutive of a commitment to Black freedom's struggle. Visual evidence captured of police by the state can institutionally propagate better visual literacy; however, it will be a visual competency in which the legal spectator is educated in how to read police attention to superficial procedural norms. There is no reason not to think that increased scrutiny of bystander and police body and dashboard footage would still support cultural amnesia of the historical opposition between Anglo-American law and the history of anti-blackness and slavery. Why would we want better visual literacy of the law's field of vision without Black freedom's struggle?

Eye-tracking is a strategy of the flesh emerging in critical approaches to analyzing police-civilian encounter videos. But we must temper these findings with the fact that the technique is part of the ascendant discourse of the algorithm that digitally constructs predictive relationships between law and the visible. Algorithmic language tools emerge in a context of ongoing struggles for liberation. But there are others. In this chapter, I have argued that the community of legal spectators in courtrooms are the real stakeholders, to whom calls for better visual competency—one that imagines the goal alongside Black liberation struggles—should be directed. In their discussion of the iconography of justice, legal scholars Judith Resnik and Dennis Curtis ask, "How might one materialize commitments to the multiplicity of vantage points legitimately to be taken into account" in matters of legal fact-finding?[51] In the "Courtwatch" movement, citizens exercise their constitutional right of assembly to *critically observe* criminal courtroom procedures. Court-watching is a form of critical observation of legal processes discussed since at least the nineteenth century in the United States.[52] The programs involve members of voluntary civic associations and traditionally include mothers, students, and retirees. Recent examples of court-watching practices build critical milieus among legal spectators to practice visual competency in and around official courtroom spaces. We might understand these assemblies as responding to the revelations of the optical fixation in implicit bias studies. For their focus is not upon the precision of what the eye sees. Rather, emphasis is on the collective situatedness of courtroom presence, whose reflections—a crucial element of the activity—are typically shared in writing, message board posts, and community gatherings after courtroom observation activities. As court-watching groups sit together, side-by-side in courtrooms across the country, their sousveillance practices are constitutive of the strategies of the flesh found at the intersection of scientific experiment and cinema.

Notes

1. See Kimberly Juanita Brown, *The Repeating Body: Slavery's Visual Resonance in the Contemporary* (Durham, NC: Duke University Press, 2015).
2. Saidiya Hartman, *Scenes of Subjection: Terror, Slavery, and Self-Making in Nineteenth Century America* (New York: Oxford University Press, 1997).
3. Kimberlé Crenshaw, "Mapping the Margins: Intersectionality, Identity Politics and Violence Against Women of Color," *Stanford Law Review* 43, no. 6 (July 1991): 1241–99.
4. Eduardo Bonilla-Silva, *Racism without Racists: Colorblind Racism and the Persistence of Racial Inequality in the United States* (Lanham, MD: Rowman and Littlefield, 2010).
5. Sora Han, *Letters of the Law: Race and the Fantasy of Colorblindness in American Law* (Stanford, CA: Stanford University Press, 2015).
6. See Safiya Umoja Noble, "Critical Surveillance Literacy in Social Media: Interrogating Black Death and Dying Online," *Black Camera* 9, no. 2 (Spring 2018): 147–160. For the relationship between contemporary bystander footage and twentieth-century lynching photos, see Erin Gray, "The Incendiary Image of Lynching: *Now!* And the Red Summer of 1965," *Black Camera, Special Issue: Contemporary Cuban Cinema* (forthcoming, Spring 2021).
7. See Roopali Mukherjee, "Bio-Work in the Blacking Factory: Police Videos and the Ethics of Seeing and Being Seen," *Black Camera* 9, no. 2 (Spring 2018): 132–46.
8. Travis M. Andrews, "The Debate around Breonna Taylor Memes: Do They Bring Attention to the Cause or Trivialize Her Death?," *Washington Post*, July 3, 2020,

https://www.washingtonpost.com/newssearch/?query=The%20debate%20around%20 Breonna%20Taylor%20memes&sort=Relevance&datefilter=All%20Since%202005. See also Allissa Richardson, *Bearing Witness While Black: African Americans, Smartphones, and the New Protest # Journalism* (New York: Oxford University Press, 2020).

9. For critical scholarship on computational algorithms, see Meredith Broussard, *Artificial Intelligence: How Computers Misunderstand the World* (Cambridge, MA: MIT Press, 2018). See also Safiya Umoja Noble, *Algorithms of Oppression: How Search Engines Reinforce Racism* (New York: New York University Press, 2018).

10. US Department of Justice Civil Rights Division Findings Regarding the Albuquerque Police Department (4/10/14) at p. 6, http://www.justice.gov/crt/about/spl/documents /apd_findings_4-10-14.pdf.

11. Charles Goodwin, "Professional Vision," *American Anthropologist* 96, no. 3 (1994): 606–33.

12. Reva B. Siegel, "In the Eyes of Law: Reflections on the Authority of Legal Discourse," in *Law's Stories: Narrative and Rhetoric in the Law*, ed. Peter Brooks and Paul Gewirtz (New Haven, CT: Yale University Press, 1996), 229.

13. Scholarship analyzing the linguistic performance of the algorithm across culture and technology include Arjun Appadurai, *Banking on Words: The Failure of Language in the Age of Derivative Finance* (Chicago: University of Chicago Press, 2016); *Fear of Small Numbers: An Essay on the Geography of Anger* (Durham, NC: Duke University Press, 2006); Andrew Culp, "Non-Constitutive Rhetoric: Or the Banality of Control" (2015), https://non.copyriot.com/non-constitutive-rhetoric-or-the-banality-of-control/; Maurizio Lazzarato, *Signs and Machines: Capitalism and the Production of Subjectivity,* trans. Joshua David Jordan (Los Angeles: Semiotext(e), 2014); Michel Callon, ed., *The Laws of the Markets* (Malden, MA: Blackwell Publishers, Sociological Review, 1998).

14. For contemporary work on the study of legal emblemata, see Peter Goodrich, *Legal Emblems and the Art of Law: Orbita Depicta as the Vision of Governance* (Cambridge: Cambridge University Press, 2013).

15. *The Brother from Another Planet*, dir. John Sayles (1984; Santa Monica, CA: MCM Home Entertainment, 2003), DVD.

16. Cultural critic Simone Browne develops the mode of sousveillance in *Dark Matters: On the Surveillance of Blackness* (Durham, NC: Duke University Press, 2015).

17. Kelli Moore, "Affective Architectures: Photographic Evidence and the Evolution of Courtroom Visuality," *Journal of Visual Culture: Special Issue Affect at the Limits of Photography* 17, no. 2, ed. Lisa Cartwright and Elizabeth Wolfson (August 2018): 208. See also Jocelyn Simonson, "The Criminal Court Audience in a Post-Trial World," *Harvard Law Review* 127, no. 8 (2015): 2174–2232.

18. Exemplary scholarship on this topic includes Dwight Conquergood, "Lethal Theater: Performance, Punishment, and the Death Penalty," *Theater Journal* 54, no. 3 (2002): 359–67; Neil Feigenson, "The Visual in Law: Some Problems for Legal Theory," *Law, Culture & the Humanities* 10, no. 1 (2014): 13–23; Neil Feigenson and Christina Spiesel, *Law on Display: The Digital Transformation of Legal Persuasion and Judgment* (New York: New York University Press, 2009); Philip Auslander, *Liveness: Performance in a Mediatized Culture* (New York: Routledge, 2008); Richard K. Sherwin, *Visualizing Law in the Age of the Digital Baroque: Arabesques and Entanglements* (New York: Routledge, 2011); Judith Resnik and Dennis E. Curtis, *Representing Justice: Invention, Controversy, and Rights in City-States and Democratic Courtrooms* (New Haven, CT: Yale University Press, 2011).

19. Moore, "Affective Architectures, 209.

20. Moore, 209.

21. Marianne Constable, *Just Silences: The Limits and Possibilities of Modern Law* (Princeton, NJ: Princeton University Press, 2005).

22. Chris M. Anson and Robert A. Schwegler, "Tracking the Mind's Eye: A New Technology for Researching Twenty-First-Century Writing and Reading Processes," *College Composition and Communication* 64, no. 1 (2012): 153. See also Andrew T. Duchowski, *Eye Tracking Methodology: Theory and Practice* (London: Springer, 2003).

23. Tanya Schneider and Steven Woolgar analyze the symbolic manipulation of scientific instruments in the nascent field of neuromarketing. Neuroscience techniques—brain imaging and other measurement technologies—are used to learn about subjects' responses to consumer products. They demonstrate how a concept of the consumer is enacted through these technologies that "simultaneously reveal and enact a particular version of the consumer that depends on achieved contrast between what appears to be the case—consumer's accounts of why they prefer certain products over others—and what can be shown to be the case as a result of the application of the [neuromarketing] technology." The statistically powerful "revelations" wrought by these tools are ironic "in the technical sense that justification for the use of these technologies depends on a constructed incongruity between what is expected and what actually is the case." Technologies of ironic revelation describe precisely the translation of eye-tracking events in perceptual bias experiments into visual representations in scientific journals where experimental conditions are typically represented as drawn models, often taking the form of cartoons or comics that feature a little homunculus seated at a computer. The subtle humor of these representations ironically refers to the experimental instrument, the footage shown to subjects. Yet a greater representational irony is afoot. The communication modeled in these studies stand in for information that cannot be represented because they are acquired surreptitiously. Optical fixations are recorded vital reports of subject's gazing behavior. Recordings of unconscious gaze patterns discover a new form of information akin to the "data-double." The significance of optical fixation is the potential social life of this data. Fixations are the product of the subject's machinic enslavement to the eye-tracker during the experiment. Optical fixations are ironic revelations of consciousness. See "Technologies of Ironic Revelation: Enacting Consumers in Neuromarkets," *Consumption Markets and Culture* 15, no. 2 (2012): 171, doi.org/10.1080/10253866.201 2.654959. See also Yannjy Yang and Chih-Chien Wang, "Trend of Using Eye Tracking Technology in Business Research," *Journal of Economics, Business and Management* 3, no. 4 (April 2015): 447–51, doi: 10.7763/joebm.2015.v3.226.

24. For feminist scholarship on techniques of transparency involving the scientific disclosure of gender and disease, see Constance Penley, *Close Encounters: Film, Feminism, and Science Fiction* (Minneapolis: University of Minnesota Press, 1991); Carol Stabile, *Feminism and the Technological Fix* (New York: St. Martin's Press, 1994); Lisa Cartwright, *Screening the Body: Tracing Medicine's Visual Culture* (Minneapolis: University of Minnesota Press, 1995). See also Paula Treichler, Lisa Cartwright, and Constance Penley, eds., *The Visible Woman: Imaging Technologies, Gender, and Science* (New York: New York University Press, 1998).

25. Ken Alder, *The Lie Detectors: The History of an American Obsession* (New York: Free Press, 2007). See also Geoffrey C. Bunn, *The Truth Machine: A Social History of the Lie Detector* (Baltimore: Johns Hopkins University Press, 2012).

26. Anson and Schwegler, "Tracking the Mind's Eye."

27. Davide Massaro, Federica Savazzi, Cinzia Di Dio, David Freedberg, Vittorio Gallese, Gabriella Gilli, and Antonella Marchetti, "When Art Moves the Eye: A Behavioral and Eye-Tracking Study," *PLoS ONE* 7 no. 5 (2012): 3, doi.org/10.1371/joural.pone.0037285.

28. R. Nagesh, "Eye Tracking Computer Control—A Review," *International Research Journal of Engineering and Technology* 2, no. 8 (November 2015): 954–58.

29. Yael Granot, Emily Balcetis, Kristin E. Schneider, and Tom R. Tyler, "Justice Is Not Blind: Visual Attention Exaggerates Effects of Group Identification on Legal Punishment," *Journal of Experimental Psychology: General* 143, no. 6 (2014): 2197. https://doi .org/10.1037/a0037893

30. Granot et al., 2205.

31. Granot et al.; Jose M. Marques and Dario Paez, "The 'Black Sheep Effect': Social Categorization, Rejection of Ingroup Deviates, and Perception of Group Variability," *European Review of Social Psychology* 5 (1994): 37–68, doi:10.1080/14792779543000011; Jan-Willem Van Prooijen and Jerome Lam, "Retributive Justice and Social Categorizations: The Perceived Fairness of Punishment Depends on Intergroup Status," *European Journal of Social Psychology* 37 (November/December 2007): 1244–55, doi:10.1002/ejsp.421.

32. Optical fixation is the value rendered by the eye-tracking apparatus, which includes the device, algorithms that transform vital data collected off the looking practice of the subject physically engaged to the apparatus. The recorded fixations occur faster than the speed of human consciousness. Paradoxically, then, the optical fixation is a unit of value assigned to the incalculable. Its calculation is made possible through Big Data techniques that unsettle the scientific method by calculating ever more infinitesimal quantities and instrumentalizing them for predictive analyses. Big Data approaches mark a shift in the history of perception from its origins beginning with the ancients who focused on perception and the senses, to optical perception conceived as the sum of added sensations, to psychological queries of gestalt meaning, to the contemporary computationally programmed vision whose features may be observed through the computer construct that is neuroimaging. Each of these approaches is organized by experimental and theoretical protocols that privilege the visual system. Similarly, the logic of Big Data appropriates from the body an endless source of data manufacturing and manufactured data. The (im)material quality of this data requires a scientific instrument, typically a networked computer, to make its appearance as visible phenomena.

33. See Judith Butler, "Endangered/Endangering: Schematic Racism and White Paranoia," in *Reading Rodney King / Reading Urban Uprising*, ed. Robert Gooding-Williams (New York: Routledge, 1993), 15–22. Further, Allen Feldman breaks through propensity toward cultural amnesia by connecting the political violence documented in the video of the Rodney King beating to the state's everyday routines of visible pain-making. The King video renders state violence in the very moment the violence of Desert Storm imagery erased the political violence upon Muslim peoples. See "On Cultural Amnesia: From Desert Storm to Rodney King," *American Ethnologist* 21, no. 2 (May 1994): 404–18. See also Brian Martin's "The Beating of Rodney King: The Dynamics of Backfire," *Critical Criminology* 13 (2005): 307–26.

34. Richard K. Sherwin, *Visualizing Law in the Age of the Digital Baroque: Arabesques and Entanglements* (New York: Routledge, 2011).

35. Sherwin, 11.

36. Peter Goodrich and Valérie Hayaert, eds., *Genealogies of Legal Vision* (New York: Routledge, 2015), 3.

37. Goodrich and Hayaert, 3.

38. Goodrich and Hayaert, 4.

39. In *A Clockwork Orange* (1971), Alex DeLarge is a juvenile delinquent who leads ultraviolent crime sprees with his gang of friends. When the Ministry eventually apprehends

him, Alex is offered a chance at rehabilitation and presumably redemption. He agrees to undergo the Ministry's Ludovico technique, consisting of a headgear apparatus that attaches restraints to the eyelids, keeping them open as a montage of violent images is played for the viewer. Alex is subjected to horrifying images set to Beethoven's Ninth Symphony. *Bladerunner* (1982) describes a future world in which commercial robots, called Replicants, illegally masquerade as human beings, sometimes unknowingly. The Voight-Kampff technique of truth is deployed forensically, consisting of a spoken interview with a subject seated before the apparatus, a small camcorder attached to a mechanical arm. Voight-Kampff combines interview questions with data collection of human eye movements. Police officer Rick Deckard pursues Replicant escapees on behalf of Tyrell Corporation. In a revealing scene, Deckard administers the test before Edmond Tyrell, the corporation's CEO, in order to test Voight-Kampff's accuracy. Narrative exposition tells us that normal interviews produce a result within twenty to thirty questions. Rachel, a subject of Voight-Kampff interrogation, requires "more than a hundred questions" from Deckard, and Tyrell confirms she is an experimental android who believes she is human.

40. Siegel, "In the Eyes of Law," 228.
41. Here readers may also be reminded of Deleuze and Guattari's idea of the "body without organs," a metaphor for a subject freed from the repressive ideological state apparatuses that organize and territorialize the self. I prefer Spillers's flesh-body distinction because it is an explicit response to the violent historical practices involved in colonialism and the process of racial enslavement, whereas the former centers its flight or detour away from repression around a colorblind reading of the human, even in its attempt at offering a critique of the same. See *A Thousand Plateaus: Capitalism and Schizophrenia* (Minneapolis: University of Minnesota Press, 1987).
42. Properly understood policing starts with slave patrols protecting the interests of plantation owners who were outnumbered by mobile slave populations. See Alex Vitale, "The Myth of Liberal Policing," *New Inquiry,* April 5, 2017, https://thenewinquiry.com/the-myth-of-liberal-policing/. See also Jill Lepore, "The Invention of the Police," *New Yorker,* July 13, 2020, https://www.newyorker.com/magazine/2020/07/20/the-invention-of-the-police.
43. Goodrich, *Legal Emblems and the Art of Law.*
44. See Goodrich, vi–xiv.
45. Goodrich, xv.
46. As Goodrich describes (xv), the emblem book "was at base a mix of military and theologico-legal inventions that rapidly set out to use the printing press to make law visible, manifest, and present in the newly enlarged public sphere, in the republic of printed letters and graven images."
47. With the term "strategies of the flesh," I am thinking of Alexander Weheliye's argument about how the enfleshments of the transatlantic slave trade do not simply impose blackness; the violence of enslavement is also generative of ways and techniques of living. See Weheliye, *Habeas Viscus: Racializing Assemblages, Biopolitics and Black Feminist Theories of the Human* (Durham, NC: Duke University Press, 2014).
48. Simone Browne, *Dark Matters,* 20.
49. Browne, 19.
50. Tavia Nyong'o, *Afro-Fabulations: The Queer Drama of Black Life* (New York: New York University Press, 2019), 212.
51. Judith Resnik and Dennis Curtis, "Epistemological Doubt and Visual Puzzles of Sight, Knowledge, and Judgment: Reflections on Clear-Sighted and Blindfolded Justices," in Peter Goodrich and Valérie Hayaert, *Genealogies of Legal Vision,* 238.

52.　See Marianne Stecich, "Keeping an Eye on the Courts: A Survey of Court Observer Programs," *Judicature* 58, no. 10 (May 1975). See also Candace McCoy and Galma Jahic, "Familiarity Breed Respect: Organizing and Studying a Courtwatch," *Justice System Journal,* 27, no. 1 (2006). See also Clare Cushman, *Courtwatchers: Eyewitness Accounts of Supreme Court History* (Lanham, MD: Rowan and Littlefield, 2011).

CHAPTER 3

Mediating Responsibility
Visualizing Bystander Participation in Sexual Violence

CARRIE A. RENTSCHLER

> There is almost *no* disagreement in our society about the
> existence of a moral requirement to rescue. The disagreement is
> only over the question of whether this acknowledged *moral* duty
> should be enforced as a legal duty.
>
> —Jeremy Waldron, "On the Road: Good Samaritans
> and Compelling Duties"

In 2012, mobile phone recordings and text messages around the kidnapping and sexual assault of a 16-year-old West Virginia girl in Steubenville, Ohio, achieved broad circulation and, in turn, catalyzed public outrage that continues to fuel major debates about the responsibility of bystanders for sexual violence. Other cases have followed, but the Steubenville case was the first to bring major public attention to "the growing use of social media content as evidence" of sexual assault, where "digital evidence" in the form of recordings and photographs could perform the role of witness to a series of assaults in which the victim herself was made unconscious.[1] In addition to picturing the assault, these forms of digital evidence made visible the participants who enacted the violence, and who contributed to additional harms against victims in their circulation and use of the recordings.

Since then, bystander participation in sexual assaults has become more visible in the press and in court cases as a result of recordings, helping to draw attention to the scale of the problem, and to the unique forms of victim harm recordings cause. In an age in which mobile phone recording is increasingly understood as a means of bearing witness to live events, recent cases of livestreamed and digitally recorded sexual violence test the very meaning of bystander responsibility for it. In current contexts of technology-facilitated sexual violence, those "uses of new technologies for the facilitation of sexual violence and sexual harassment," bystander responsibility is reimagined around key cases where bystanders participate in the recording and dissemination of acts of sexual violence.[2] In some of the cases examined here, videos in which participants recorded sexual violence were made public through the efforts of hackers such as KnightSec, an offshoot of the Anonymous hacker network, and a Canadian-based network of hackers also identified with Anonymous.[3] These videos and other "digital witnesses" to bystander participation reshape how activists, legal scholars, and others conceive of bystander responsibility for sexual violence as something that exceeds, and calls into question, existing legal frameworks on bystander culpability for sexual violence, violence where, as Kelly Oliver argues, "recording rather than reporting may become a new standard for prosecuting rape cases."[4]

By responsibility, I mean a state that emerges as a consequence of being "subject to the unwilled address of the other" that constitutes a way of receiving and acting on the scenes of violent subjection and the ways one is called to act in the face of another's violation.[5] To be responsible means to become a witness to and of other people's harms and needs, someone who can enact their capacity to respond and be accountable in the face of a non-role-defined obligation to care for another person in need, without necessarily *wanting to,* or *deliberately choosing to be,* responsible. In this chapter, I refer to people who record or livestream sexual violence at different points as participants, bystanders and digital witnesses, to emphasize the different acts of participating in, standing by and taking account of sexual violence.

I am particularly interested in how contemporary feminist frameworks on sexual violence understand bystanding as a capacity for acting against and intervening in it. Drawing from philosopher Kelly Oliver's use of the term "response-ability," I consider how people's abilities to respond to sexual violence using the tools of Internet-connected mobile devices create highly contingent conditions in which responsibility can be but is not always practiced.[6] Having the capacity to respond to sexual violence does not guarantee one will bear and take responsibility for it. To perform responsibility requires that bystanders orient themselves and their concern toward care for victims, where they recognize and understand the harms they suffer. It also requires disidentifying with perpetrators and the harms they cause. As I argue elsewhere, "The bystander position is always already one of participation that comes with consequences."[7] These consequences are not determined in advance, nor are they always easily categorized as perpetrator-supportive or victim-supportive. From the perspective of participants, then, recording can both participate in *and* report sexual violence. In this way, what bystanders do or do not do in the contexts of sexual violence has one of two effects: it either aids victims or it causes them harm. There is little middle ground here: you either help or you are part of the problem.

Recent cases of sexual violence that have been digitally recorded and/or livestreamed reveal the connections between *recording, enacting*, and *being responsible for* sexual violence. These cases often involve more than one attacker, like the one in Steubenville, Ohio. Those who record and livestream sexual violence are increasingly visible as complicit agents in violence in ways with which the law has yet to fully grapple. Some forms of sexual violence are enacted *in order to be* recorded, so that recordings can be shared among participants and their larger friendship networks via social media.[8] The social media dissemination of sexual assault recordings extends the harms that assailant(s) commit against the victim(s), creating new harms as they shame victims, expose their violation to a broader public, and take voyeuristic pleasure in the infliction of violence against them.[9] Even incidental recordings of sexual assault become tools for harming victims when they are shared among social networks. As a report in the *Guardian* explains, "It is staggeringly routine for a rape, especially if it occurs in a social setting, to be accompanied by some form of digital trophy-taking . . . not evidence, but a memento."[10]

Internet-connected mobile phone cameras can thus become tools of perpetration. Legal scholars Holly Jeanine Boux and Courtenay Daum explain that recording adds a "new dimension to these crimes" of sexual violence, where violence that is digitally recorded, re-spectated and shared with others more broadly enacts gendered, sexualized, and racialized forms of violation and subjugation against victims that extend beyond the event of the assault itself.[11] In the context of sexual assaults in which the victim is incapacitated and/or unconscious, like the Steubenville case, video recordings become trophies that perpetrators and other participants can re-watch and share with other viewers.[12] Such videos represent a kind of "perpetrator-cam" perspective, analogous to video trophies kept by soldiers engaged in war crimes and other forms of violence.[13] Recording-enabled, Internet-connected mobile devices are not, then, simply present in the context of sexual violence, particularly when there is more than one attacker. They become technologies of assault, defining the kinds of sexual violence that are enacted against victims who are often unconscious, unresponsive, and highly inebriated, and who therefore cannot possibly defend themselves.[14]

By publicizing these recordings, news coverage and anti-violence movement actors target bystanders as agents who bear responsibility for sexual violence and its victims. Contemporary anti-rape activists construct bystanders not only as accomplices to perpetrators, but also as potential interveners who can call out, disrupt, and stop sexual violence and the cultural practices that support it.[15] Today university and high school bystander intervention programs and US federal Title IX policy identify the bystander as someone who can prevent sexual violence around their proximity to it and their location in community. As peers, friends, co-workers, and other members in community, bystanders often witness sexual violence and the cultural practices, behaviors, and beliefs that support and legitimate it. These forms of witnessing are also increasingly happening online, where respondents to Pew Research Center surveys in the United States report high rates of witnessing gender, sexual, and racial harassment of others in addition to forms of threatening and sexually violent behavior.[16] Because of their social proximity to sexual violence and its social conditions in a networked media environment, they are uniquely situated to interrupt many of the behaviors located along the continuum of sexual violence: from verbal harassment, cat calls, and rape jokes to non-consensual touching, sexual coercion, and sexual assault, among other acts and behaviors.[17]

In this moment, as this chapter demonstrates, holding bystanders responsible for sexual violence aims to rectify the "justice gaps" in criminal justice approaches to sexual violence, those "dramatic gaps between the number of offences recorded by police and the number of convictions" for sexual assault that Jennifer Temkin and Barbara Krahé identify.[18] Each year in Canada, where I am located, less than 1% of the sexual assaults that occur result in *any* kind of legal sanction.[19] Such gaps exist in part because of rape myths that operate in the courts and legal system, through which members of the legal professions misunderstand, mistreat, and mistrust the survivors of sexual violence.[20] As Temkin and Krahé explain: "Judgements about sexual assault are skewed in the direction of low conviction rates partly because of the widely held attitudes about rape which undermine the position of the complainant and benefit the defendant."[21]

As a response to these justice gaps, survivors of sexual violence and their advocates create *other* models of justice that aim to hold perpetrators accountable through means other than criminal justice. They also aim to hold accountable a broader community of participants in sexual violence, such as those who help enable the violence, those who support it while it was happening, those who spectate it, and those who do nothing to stop it. Holding bystanders accountable is, for some, part of an alternative model of justice for survivors that avoids, and challenges, criminal justice as a framework of redress. These models emerge from both feminist-of-color prison abolition movements and transformative justice movements that seek to dismantle the links between the systemic violence of the criminal justice system and the systemic structures of intersectional gender and sexual violence.[22] By drawing on non-carceral frameworks of transformative justice—and in lieu of further incarceration of those who commit sexual violence—they create forms of justice based in support for survivors and shared community accountability *to* survivors and their needs, and *for* those who do harm.

Within Anglo-American legal frameworks, bystanders bear no legal "duty to rescue" those being harmed, unless their role responsibilities vis-à-vis the victim obligate them to, as most "Good Samaritan" laws articulate.[23] Most legal duties to rescue are specific to relationships in which one person bears role responsibility for the safety and well-being of another as a consequence of their relationship to one another, such as a parent to a child, a doctor to a patient, a teacher to a student, or a spouse to another spouse. Except in very particular cases around medical emergencies and car accidents, US and Canadian law tends not to ascribe legal responsibility onto those who could, by their proximity or ability—often *as* bystanders—offer aid to others in emergencies. As one legal scholar explained, during the height of debates in US and Canadian legal communities about the potential usefulness of Good Samaritan legislation in the 1970s and 1980s: "For the most part, neither our tort nor our criminal law concern themselves with the good that one doesn't do, but rather attempt merely to forbid doing harm."[24] Nonaction, in other words, is not considered a kind of action around which individuals are typically held civilly liable or criminally responsible. As legal scholar Kathleen Ridolfi explains, there is also no legal duty to rescue in cases of sexual assault. While there should be, she argues, "little" still "is owed by way of a duty to rescue" in Anglo-American law for those targeted by sexual violence.[25] In response, advocates and others turn to different means to impel, encourage, and demand bystander responsibility for sexual violence.

This chapter examines how efforts to impel bystander responsibility deploy photography and digital video *of* bystanders as part of a larger public demand to hold them responsible. In press reports and online platforms, bystanders are made visible through some of the same media that implicate bystanders in sexual violence—in surveillance video footage, courtroom footage, and press recordings. Advocates and others use images and recordings of bystanders to sexual violence in a way similar to how photographs of German bystanders who witnessed the Nazi concentration camps were circulated in the press and used to make claims for the responsibility bystanders shared for the Holocaust: they make bystander complicity visible in order to call for bystanders to be held responsible for what they witness.[26] Picturing bystanders in these ways attempt to do "rape justice beyond the criminal law" as part of a larger redefinition of what justice is for sexual violence survivors.[27] Following feminist criminologists, I examine what shared responsibility for sexual violence looks like when justice "increasingly appears as technosocial practice"—where justice is enacted through the same means, and media environments, in which technology-facilitated sexual violence is committed.[28]

To do so, I examine a set of cases of sexual violence that were recorded or livestreamed and spectated by others. Each case appeared in court, in the press, and in contexts of social media distribution and discussion, where some of the harms were enacted against victims, and where discussants, in turn, aimed to connect bystanders' answerability with their agency in the violence.[29] I first examine how participants in the Steubenville assault recording were imaged in *The Atlantic* in an effort to make them responsible. I then turn to the case of Marina Lonina, an eighteen-year-old Ohio woman who livestreamed the sexual assault of her friend, reportedly in an effort to get help from someone who might be watching the stream. In this case, someone out of state who followed Lonina's Periscope stream saw the broadcast and called police to report the assault. Lonina was brought up on several charges, including the charge of sexual assault, for the 2016 livestream; she eventually accepted a plea deal for the charge of obstruction of justice. I then examine a high-profile case from Nova Scotia, Canada, where two male teenagers faced conviction as a result of their video documentation of their sexual assault of fifteen-year-old Rehtaeh Parsons in 2011. The two young men distributed the video to students at Parsons's high school and, after her cyber-harassment and suicide, were brought up on child pornography charges in 2013. The chapter concludes with a brief examination of the case of the bystanders who intervened into Brock Turner's sexual assault of Chanel Miller at Stanford University in 2015, and the ways they have become visible as responsible (male) agents of intervention against Turner's—and his family members'—failure to take responsibility for his assault.

Each case brings the interrelation of law and digital recordings into view as "technologies of responsibility" by making bystanders visible as participants in cases of sexual violence. Picturing bystander/participants in digital photographs and surveillance video footage becomes a key means to create bystander responsibility for this violence where it otherwise might not exist, for instance, in criminal law. Their visibility, as we will see in the cases, comes without any guarantees that bystanders will effectively be held responsible, or that institutions of justice will take up the call to make them responsible. To enact effective accountability processes in which bystanders can and do take responsibility for sexual violence requires that justice protocols—and scripts—are put in place, whether it is the context of law or transformative justice responses. As Martha Merrill Umphrey argues, law is performative, particularly in the context of trials, where an iterative and interpellative process enunciates, produces, and *does* the law by calling legal subjects, and legal violations, into being.[30] This requires scripts for calling bystanders into being as agents who can be held responsible for sexual violence, but who also can do the work of enacting their accountability for it.

Bystander videos of sexual violence constitute part of this performative work that law does, informing legal conceptions of bystander responsibility for sexual violence. But they also make bystander responsibility visible beyond the contexts of criminal law and trial courts, in social media networks and in popular press coverage that reveal the lack of legal scripts and protocols for adequately holding bystanders responsible for their roles in sexual violence. The need to make bystander responsibility visible is a response to the absence of such scripts and protocols. When those legal and extra-legal protocols exist, they can do some of the work that those acts of making visible bystander responsibility—and failure—demands.

In each case examined here, the identified individuals became the subjects of high-profile news coverage and social media discourse that sought—sometimes successfully—to make them responsible for the violence by distributing recordings of them on security camera footage, in the case of the young men charged for Parsons's assault, and in videos taken from the courtroom, in the case of Lonina. As these cases suggest, to make participants in recordings of sexual violence responsible, bystander irresponsibility is "called out" by making it publicly visible on a larger scale: through digital technologies of recording, playback, and distribution. Around those recordings, activists and others build scripts for bystander responsibility that draw from feminist scripts of sexual assault resistance and bystander intervention: those frameworks that help people learn how to intervene in sexual violence and sexual harassment and how to communicate and enact that intervention into the "cultural scaffoldings of sexual violence."[31]

Picturing the Agents Who Enact and Record Sexual Assault

In the 2012 case in Steubenville, Ohio, videos taken by accomplices to the victim's kidnapping and sexual assault while she was unconscious helped to convict two young men, Trent Mays and Ma'lik Richmond. Both received one-year sentences for digital penetration of the victim while Mays received an additional year for disseminating a pornographic picture. Three adults were also indicted for obstructing justice and evidence tampering for their role in trying to suppress the case. The prosecution also attempted, but failed, to bring charges against the witnesses and social media bystanders for the role they played in the violence committed against this young woman, backed by a body of evidence from text messages, and a twelve-minute video taken on the night of the assault, in which at least six people who participated in and were present during the violence talk and joke about the assault and the victims' state of unconsciousness. Her body is visible in several of the video's shots, wrapped in a blanket on the floor by the armchair on which one of the video's key subjects, Michael Nodianos, sits. Even as videos like this one evidence the participation of a group of people in the violence committed against the victim, their status as legally evidentiary digital witnesses to bystander participation carry no guarantees for prosecution. A review of the 2019 documentary *Roll Red Roll* about the case explains these limits, where even "callous texts and vicious comments, like the twelve-minute video of one heartless boy making 'dead girl' jokes about his friends' sexual assault *aren't even illegal*, as lead special prosecutor Marianne Hemmeter acknowledges with a weariness that you can tell comes from experience."[32]

The video camera, and the recording itself, both serve as digital witnesses to the assault and its larger scene of perpetration and victim degradation over time. While at least six other people are visible or audible, the video centers on Michael Nodianos, his body, and the jokes he tells, helping identify him as an individual who could be held accountable. The video was posted in full to *Deadspin.com* in January 2013 after KnightSec hacked into participants' phones and the high school website and then released the video to the public.[33] On audio, participants can be heard commenting on the victim, other participants, and their behaviors. Listening to the video, it becomes clear that several people are laughing and groaning at the inappropriateness of the jokes Nodianos tells: he has an appreciative audience to whom he provides more and more degrading material. Press reports focused on the jokes Nodianos made over how "dead" the victim is, as she lays next to him unconscious, and "how raped" she is, engaging in what Kelly Oliver calls the sexual "fantasy of the dead girl."[34] His jokes all begin with scripted statements such as "They raped her more than . . . (fill in the blank)," followed by answers such as "the Duke Lacrosse team," a reference to a sexual assault report made against the team. Another joke script begins, "She's so dead that she . . . (fill in the blank)," to which he answers, at one point, "Trayvon Martin," in reference to the seventeen-year-old victim shot dead by George Zimmerman in February that year. At several points, Nodianos laughs so hard that he cannot speak, suggesting that, like misogynist online trolling cultures, "it is easy to laugh ironically . . . when your particular body is not in danger."[35] The recordings Nodianos and his peers made of the sexual assault that night were part and parcel of the online and offline misogyny, and seemingly ironic and detached violent humor, that young men use against young women in Internet culture.

While videos like this one reveal the participation of a number of bystanders, including at least one teenage girl, the press used the video to focus bystander responsibility around him. Still photos taken from the recording show Nodianos's laughing face, his eyes squeezed shut, his mouth wide open, and his cheeks flushed. *Deadspin.com* used a still of Nodianos's face in the story that accompanied the video's release, juxtaposing it with a digital photograph of Richmond and Mays carrying the victim unconscious by her hands and feet, directly associating his joking about the victim being assaulted and unconscious with her sexual assault.[36] While it is somewhat difficult to tell from the poor audio quality of the cellphone video, one or two participants challenge Nodianos and the others for their assaultive behaviors. One person can be heard saying, "Dude, you might go to jail." At another point soon thereafter, he says, "This is not, like, funny here. What if that were your daughter?" while another person asks, "What if that happened to my little sister, who just turned sixteen?" reflecting on the fact that his sister could be targeted for the same kind of violence by a group just like theirs. Nodianos is then asked: "Do you even hear what you are saying? You are talking about her being dead. You are sick. This girl is drunk."

To listen to the whole video is to hear a somewhat more complex, but also quite harrowing, performance of mostly young men's homosocial participation in sexual violence as bystanders. Not everyone agreed with what was happening, but they did not intervene to stop it or to get help for the unconscious young woman. Some people heard on the video seem to recognize that what they are doing is harmful and degrading to the victim, for instance, when one person describes how someone has urinated on the victim.[37] Their recognition of the violence they participate in does not, however, lead to doing something about it. The video thus reveals both a homosocial celebration of sexual violence against an unconscious girl as it simultaneously records the attempts some participants make to enact the barest form of response-ability: they respond to the violence as violence, and to Nodianos and others as responsible to and for it. They implore them to think about the person they are harming, and they tell Nodianos that he might be arrested by the police. They also express empathy toward the victim by imagining the young woman as someone they know personally, even if not knowing someone personally should not make them less worthy of care and defense.[38] While the video implicates each of the participants in the harms of the assault and the harms of the recording, then, it also reveals the ways some participants judge the behavior they witness in the moment.[39] They do not, however, talk about the role they play in being an audience to Nodianos's behaviors, or having been, perhaps, present during Richmond and Mays's sexual assault of the victim.

Photos of the unconscious victim, which broadly circulated in the wake of the assault, represent a particular kind of "image exploitation," where "victims view their own lifeless bodies being dragged, dropped, violated, abused and raped, not as participants in the scene but as observers of it."[40] Such exploitation is "a distinct form of sexual abuse, involving the non-consensual creation, possession, or distribution of an image or images depicting the victim as nude, semi-nude, engaged in consensual sexual activity, or being sexually assaulted" in the form of "a photograph, screenshot or video recording."[41] By circulating the image via "cellphones, email, social media, and the Internet, an offender can distribute photographs and video to the victims' circle of friends, family, and colleagues, as well as countless denizens of cyberspace" in an attempt to expose and shame the victim in front of those they know.[42] As a concept, then, image exploitation gives a name to the role spectators of sexual assault videos play in causing harms to the victims, harms that can be legally identified and adjudicated.

While *Gawker* and other media exposed the voyeuristic and seemingly joyful pleasure some participants took in Jane Doe's sexual assault and image exploitation, *The Atlantic* revealed the attempts some of these participants made to explain themselves after the fact. In video stills taken from surveillance video footage (and broadcast on *ABC World News Tonight*) of individuals involved in the assault as they are questioned by local police, *The Atlantic* story pictures their seeming remorse, *as if* they might be preparing to take responsibility for their actions. Some explain how they failed to apprehend what was going on during the assault, while others state they did not know what actions they could have taken to stop it.[43] These explanations are fairly typical scripts used by bystanders to account for why they did not intervene into a situation that was, or was becoming, sexually violent.

The Atlantic and *Gawker* stories attempt to model a process by which the case to hold bystanders accountable for their participation in sexual violence is made by making them visible as complicit, and sometimes remorseful, subjects. By performing remorse, they begin to recognize and reckon with the harms in which they participated. Photos and video stills taken from a surveillance camera in the room in which they were questioned by police visually identify them as accomplices to the violence who, otherwise, would not have been knowable as agents who could be held accountable (except, perhaps, by those people who were locally involved in the crime's commission and the subsequent police investigation).[44]

Some legal professionals train prosecutors and lawyers in how to bring charges and build cases against those engaged in image exploitation through the recording of sexual assaults, connecting legal and extra-legal responses to bystander participation. Jane Anderson and Supriya Prasad advise the legal community to see bystanding not as a neutral position but instead as forms of complicit, *liable* involvement in sexual violence:

> Often dismissed as simply bystanders, individuals who film an assault may have orchestrated the assault or identified and encouraged perpetrators to engage in the assault—sometimes for the purpose of creating exploitive images. According to the theory of accomplice liability, where offenders are acting in concert, aiding, or abetting, each participant is held accountable for each other's actions. Similarly, conspiracy charges are applicable where the evidence establishes an explicit or tacit agreement to commit the assault and [in some cases where] the overt act of filming the assault is completed. Accomplice and conspiracy theories of prosecution are strengthened by the fact that the image the "bystander" created can be viewed as a lasting "trophy" of the crime itself.[45]

Here, the "bystander position" is one both identified with, and a participant in, the perpetration of sexual violence. To be made legally legible, bystander complicity must also become knowable according to the harms it produces. To do so requires making bystander complicity culturally visible for both legal and extra-legal forms of responsibility taking.

Legal scholar Renu Mandhane further argues:

> A lack of culpability for bystanders fails to capture the reality of the situation for
> sexual assault victims. It is probable that the presence of bystanders causes fur-
> ther psychological harm to the victim. Yet, the law allows bystanders to watch
> as a woman is dehumanized free from any obligation to help the victim.[46]

Mandhane argues that legal conceptions of liberty should be based in the
vulnerabilities women face for assault that "would recognize *in law* the
relationships we all share with one another and the rights that result from
these interconnections," including the relationships between bystander
and victim.[47]

As Anderson and Prasad suggest, prosecutors can be trained to better understand the harms of image exploitation. They translate feminist
frameworks on image-based gender violence, which aim to hold bystanders
responsible as complicit agents in its perpetration, into criminal justice
strategies against those who record and share digital recordings of sexual violence.[48] Image exploitation works by denying victims control over
the representation of their often-unconscious bodies while they are being
violated. And, as Anderson and Prasad note, when others post and share
images of sexual assault online, they become part of the public domain:
no longer private, personal "trophies" of assailants and their friends, but
easily searchable by anyone online, where the "crime goes viral."[49]

As feminist philosopher Cressida Heyes describes, the networked sociality
of bystander culture constitutes "a community of voyeurs created around the
images" of sexual assault that they and others create, where they can "pore
over their deeds or brag to their friends."[50] Their modes of homosocial networked spectatorship enables perpetrators, accomplices, and other complicit
subjects to bond through their shared spectatorship of the recordings.[51] The
video's subjects, on the other hands, are often "forced to see their bodies as
living corpses through the eyes of witnesses who claim they look 'dead' and
'lifeless.'"[52] In these contexts, as Kimberly Allen suggests, law has tended to
"ignore the motivating role that 'spectators' play in gang rapes."[53]

Performing Response-Ability in the Context of
Social Media Attention Economies

In 2016, eighteen-year-old Marina Lonina was brought up on criminal
charges in the state of Ohio for recording her seventeen-year-old female
friend's sexual assault by twenty-nine-year-old Raymond Gates, a man the
two had met the night before the assault. On the night in question, Lonina

livestreamed the assault via the Periscope mobile phone app, which is owned by Twitter, as the *New York Times* reported, "in hopes that someone might come to the rescue of her friend."[54] By viewing the livestream, someone from out of state *did* call the police, which suggests that the livestream succeeded in getting help from someone not in the room with them; at the same time, the broadcast of the assault also did great harm to the victim by showing her being assaulted to an audience of followers that responded with streams of "like" emojis. As reported in an Associated Press story, the victim stated that Lonina had not tried to get help but in fact had set up the sexual assault by Gates.[55] The press also reported that the day prior to the assault, Lonina had posted a nude photo of her friend online; she was also brought up on felony charges for doing so.

This case, in addition to the others discussed here, catalyzed public calls in the United States to make onlookers legally responsible for the violence they witness, representing what one journalist described as "an appetite to punish not only the rapists, but anyone who facilitated the crime."[56] After being charged with counts of rape, sexual battery, kidnapping, and pandering sexual matter, Lonina pled to a lesser charge of obstruction of justice, where she "admitted that she failed to report the rape and did not submit evidence from the incident."[57] She was sentenced to nine months in jail.[58]

The gendered framing of her actions, and presumed guilt, are central to the reporting on this case and the ways she used livestreaming capabilities to both participate in the violence and, seemingly, seek help. A couple of things become clear from these press reports: first, Lonina is understood to be livestreaming for attention, as part of an online economy in which girls and young women, in particular, are encouraged to develop a self-branded identity that can be performed via social media posts and uploads of selfie photos and videos and then exchanged for attention, reputational merit, and "followers."[59] This economy connects mass media and social media networks of reporting to that of status updates and the forms of social media accounting people use to comment on their day-to-day life.[60]

"Live video is the new selfie," an NPR radio report on the case declared.[61] As prosecutor Ron O'Brien explained, while he watched the ten-minute recording of the live stream, "likes" were "popping throughout the whole thing."[62] O'Brien also remarked that Lonina could be heard "giggling and laughing" on the livestream audio, sounding like a teen girl performing for the kinds of attention from followers that structure the economies of social media posting and sharing.[63] Lonina had reported that she got "caught up in the likes" as a way of explaining why she kept livestreaming the assault, a statement that reveals both how plausible and simultaneously horrific the behavior is. Others represented her as a callous "bad" or "sick" girl (as the UK *Mirror* described her) whose actions of recording her friend's assault were found to be all the more reprehensible since, had she really wanted to help her friend, she could have used her phone to call the police or texted someone to get help instead.[64]

It is certainly true that Lonina could have simply called the police herself. But the fact that she didn't also suggests that there is another perhaps more central way that she and others see their mobile phones as their access to safety and help in an emergency. As reported by the NBC affiliate WCMH in nearby Columbus, Lonina's lawyer Sam Shamansky explained that his client had in fact "tried to help the victim by asking her followers, 'What should I do now?'"[65] By livestreaming, an "audience responds in real time, typing comments, asking questions, or tapping the screen to send a stream of animated hearts as a form of appreciation."[66]

In this context, Lonina's livestream was not an unusual use of livestreaming capabilities, similar to how some injured people inside the Bataclan music club in Paris on the evening it was attacked by terrorists in November 2015 used Facebook livestreaming capabilities to encourage the police to raid the building.[67] Individuals used Periscope and Facebook Live to broadcast from the scenes of several other terrorist attacks and mass shootings; in response to livestreams of a 2016 terrorist attack in Munich, Germany, police even set up an online portal through which bystanders could submit their recordings to aid the police investigation.[68] "Mobile live-streaming apps," then, can "bypass traditional institutional structures and provide citizens with direct, immediate access to global audiences and viewpoints."[69] They can also, circuitously, speak to police.

Lonina *could* have been both asking for help with her livestream and "getting caught up in the likes" of responding audiences. What does not get talked about in this case are the likes that viewers expressed in response to the livestream. "Likes" are notoriously hard to interpret as the polysemic symbols they are on social media platforms. Social media attention economies presume that people live their lives in networked, multiplatform media environments. As Lee Humphreys explains, "It is because the everyday is being recorded regularly that the eventful moments can be captured."[70]

If we are to study the Arab Spring, for example, through Twitter, then mobile phone adoption and use—the mode through which most people access Twitter—had to be in place before the elections and demonstrations. If people are to capture police abuse on their camera phones, like they did in the case of Eric Garner, who died of a heart attack while New York City police were trying to arrest him in 2014, then people must have had them in their pockets to begin with.[71]

For Humphreys, the act of recording and streaming acts of violence with mobile phones comes from a self-referential capacity people develop to document their lives to others using phone cameras and social media platforms. As Paul Frosh explains, the "gestural image" of selfie photos and videos allow people to "see me showing you me."[72] Recording extraordinary events becomes an extension of the ordinary habits of using social media to record one's daily life.

Livestreaming also constitutes part of a social media ecology where people are more likely to turn to their mobile phones to record and livestream an extraordinary event—including a sexual crime—rather than responding in other ways, such as by calling police. And, as contemporary notions of social change are increasingly connected to forms of participation in social media attention economies, livestreaming represents a form of "citizen broadcasting" that makes up part of this attention economy, connecting one's "followers" on social media platforms to social issues and everyday occurrences via easy-to-access apps on one's mobile phone.[73] In many cases of recorded and livestreamed police violence such as the police shooting of Oscar Grant by BART transit police in Oakland in 2009 to the choking death by police of Eric Garner in New York City in 2014 and the suffocation of George Floyd by Minneapolis police in 2020, it is (often) not safe for racialized bystanders to intervene in the violence by calling the police when police are the assailants.

As an avid social media user, Lonina would have been aware of the association between livestreaming and efforts to stop violence and make social change. Within the context of efforts to draw attention to social problems within social media environments designed for self-branding and the accounting of daily life, Lonina's broadcast of her friend's sexual assault could look like, and even be, *both* an attempt to get help *and* a performance of her participation in the violence. Very little reporting on the case held open this dual role that Lonina's live broadcast could have played. One rare story by *Vice* magazine opined on how "the bizarre case begged the question of whether or not Lonina was responsible for the same sex crimes as the older, male assailant, despite apparently playing no direct role in the assault herself."[74] Another report characterized the case rather dismissively as "the latest American saga pitting law enforcement against apparently social-media obsessed teenagers."[75] That same story in *Vice* magazine asked directly: "Does livestreaming a rape make you a rapist?"[76]

The ways in which Lonina is made visible, however, as part of the legal case against her for livestreaming, instead works to identify her as a witness that is both culpable and complicit in the sexual violence. As a *New York*

Times video captures the defense counsel's statement at the time of Marina Lonina's initial entry of her not-guilty plea, viewers watch Lonina standing by her attorney's side, chewing her lip, and casting her gaze downward, the image of a young girl caught doing something she should not have. She wears an orange jumpsuit, her long blond hair cascading in stringy wisps around her face and shoulders. Lonina looks drained. She wears the haggard face of someone who has been made to bear (some) responsibility for her friend's assault, and someone who has spent some time in county jail.[77]

As Sharon Sliwinski suggests in her history of human rights photography, "Justice must not only be done, it must be *seen* being done."[78] The visibility and virality of bystander videos and livestreams of sexual violence suggest that injustice committed by bystanders needs to be seen as part of the process of seeking justice for what they have done—or not done. The case of Nova Scotian Rehtaeh Parsons's assailants reveals how the inability and refusal to picture perpetrators as key witnesses to their own violence protects sexual assailants from full legal responsibility for the harms they created.

The Effacement of Perpetrator Responsibility in Pictures

In August 2013, two young men were brought up on charges of making and distributing child pornography images of Rehtaeh Parsons, a fifteen-year-old who eventually reported being assaulted by the boys and two other young men while she was inebriated to the point of becoming sick on a November night in 2011.[79] Child pornography charges were brought against the young men two years after Parsons and her parents reported her sexual assault by the four young men in a house in Cole Harbour, Nova Scotia. This was four months after Parsons killed herself as a result of the assault and subsequent cyberbullying she suffered after a digital photograph of her assault was circulated to students in her school.[80] For Nova Scotia Provincial Court Judge Gregory Lenehan, the photograph was "the domino" that "started the cascading events that led to her death"—a weapon used to cause a number of additional harms to that of the sexual assault.[81] Journalists and editors asked why no one else intervened to help Parsons, both at the time of the reported assault and later, when the photograph of an individual assaulting her was broadly circulated among students at her school. As Charlie Gillis, managing editor of *Maclean's* magazine, stated, "No investigation seems likely to answer another, far-reaching question arising from [Rehtaeh] Parsons' death: when the pictures first emerged, why did none of her peers speak up?"[82]

In the photograph of Parsons that was circulated, she is pictured vomiting out a bedroom window while a sixteen-year-old young man sexually assaults her, smiling to the camera and making a thumbs-up sign.[83] As the young man stated in an interview: "I just posed for the picture and he [his friend] took his cellphone out and took it."[84] The young man pictured was sentenced to a year of probation for distributing child pornography; the other pled guilty to the production of child pornography for taking the photograph and received a conditional discharge with a year's probation and a suggestion that he get counseling and stop drinking.[85]

The latter also performed a highly limited version of taking responsibility tied to his specific role in circulating the images, and not for the harms done to Rehtaeh Parsons as a victim of sexual assault. As the young man who took the photograph stated at his sentencing, "I have plead[ed] guilty to distributing child pornography, not a sexual assault," and "I will not live with the guilt of someone passing away, but I will live with the guilt of sending the picture." What counts as "responsibility" here is contained around the legal specification of unique and separate acts of image exploitation.

Photographs of the young men entering and leaving the courtroom broadcast on Canadian television and then published in the British press partially circumvented the publication ban on identifying them by blurring their faces. Both appear in individual photographs in a courthouse stairwell, their faces and hair made completely unrecognizable. The young men are both pictured, and not pictured, as responsible agents that readers and viewers are unable to look in the face.

In a trend we increasingly see in cases of sexual assault where there are digital video records and/or photographs, some fail to see what non-consent looks like, seeing "sex" rather than the conditions that make continuous and affirmative consent difficult if not impossible to enact.[86] As Kelly Oliver suggests, recordings of sexual violation committed against unconscious young women often celebrate the conditions of non-consent.[87] Some journalists and police representatives described what the images in the photo of Parsons portrayed as "sex" rather than violence or non-consent. One Canadian columnist referred to it as "sexually graphic," while police termed it, at various times, an "inappropriate image" rather than an image of sexual assault.[88] In a 2013 article, Christie Blatchford, a high-profile Canadian columnist who was known for her anti-feminist stances on cases of gender violence, went so far as to describe the photograph of Parsons as a potential image of consensual sex, one that "shows virtually nothing that would stand up in court" as evidence of sexual violence. She further describes the image in almost playful terms, depicting "a male naked from the waist down, giving a thumbs-up sign, *pressing into the bare behind of another person who is leaning out a window.*"[89] Rather than focus on the inebriated and unwell state of the young woman who is hanging out of the window, and in no condition to provide sober consent, Blatchford describes the "fun" the young men appear to be having, and which they confessed they were having. "If there even was a sexual assault going on," Blatchford suggests, the images cannot be trusted to prove it.[90] The forms of "digital witness" such photographs offer—as records of sexual violence—thus "reveal the continued shortcomings of the criminal justice system and the persistence of entrenched stereotypes about sexual violence" that rely upon *not* seeing sexual violence in recorded depictions of it.[91]

In Canada, there is a legal apparatus in place for criminalizing the making and distribution of sexually exploitative images that defines the makers and distributors of these images as responsible agents. Canadian lawmaking pays particular attention to the harms and injuries done to victims when digital videos and photos of their assaulted bodies circulate on social media networks, articulating a context of legal responsibility for those actions. Several Canadian cases motivated these recent laws, including that of Rehtaeh Parsons and of fifteen-year-old Amanda Todd (of British Columbia), who also committed suicide after experiencing intense cyber harassment when her adult male harasser posted, and peer bystanders digitally circulated, partially nude pictures of her. Bill C-13 "Protecting Canadians from Online Crime Act" was passed in 2015 and revised Canada's Criminal Code to include the new "offense of non-consensual distribution of intimate images." Nova Scotia's 2013 "Cyber Safety Act" was a direct response to the Rehtaeh Parsons case and was the first Canadian provincial legislation of its kind.[92] Bill C-13 also gave Canadian law enforcement agencies and security services increasing latitude to surveil the electronic communications of Canadian residents and citizens without a judicial warrant, directly linking the new legal mechanism to protect victims of cyber violence to state interests in population surveillance as a means to address perceived threats of domestic and international terrorism.[93]

Rather than legislate against failures to intervene to stop an assault—as a duty to rescue likely would—lawmaking around witness recordings of sexual assault defines it as a crime of image-based exploitation. Rather than offer a fix for the problem of sexual violence that is also digitally recorded, these laws instead provide a set of what Scott Leitch calls "normative devices and practices" that organize and distribute responsibilities across the spectrum of acts, and actors, that constitute sexual violence as an event to be watched live, recorded, re-watched and re-viewed. These laws become tools for making witnessing agents responsible for what they record rather than holding them responsible for the sexual violence that is being recorded, or, in some other cases, failing to assist a person who is being assaulted. They seek to criminalize the recording and photographing of sexual violence, to make bystanders responsible as documentarians of sexual acts, but often not as participants in sexual violence. This suggests that there are hard limits to visibility as a means of performing and enforcing bystander responsibility in criminal justice contexts.

Conclusion

Bystander videos, like other forms of media witnessing, "put society on view to itself."[94] Like many examples of visible bystander complicity in sexual violence, the cases examined here deploy an audio-visual logic in which taking responsibility remains unfulfilled. As I argue elsewhere, "Bystander videos dramatize the contingencies, insecurities and fallibility of bystander response-ability. In many cases, the bystander experience may ultimately fail to deliver on this capacity to respond in an interventionist way."[95] While law and images of bystander participation in sexual violence are meant to work as technologies of responsibility, we see how constrained and context dependent those practices are.

Responsibility is a set of practices that must be performed, as legal theorist Scott Veitch argues.[96] To take responsibility as a bystander is to perform becoming a responsible agent in ways that are visible and audible to others, both in the courtroom and in the networked public spheres of the press and social media reporting.[97] In this way, taking responsibility is a way of embodying what it means to bear witness to sexual violence as an agent who recorded it. When bystanders are made visible as the subjects behind the act of bystander recordings—either when they appear in court, in court documents, or in video or audio form—it is in order to enforce a mode of "performing responsibility" that bystander subjects must enact. Some fail at this performance; others succeed.

As I have argued, then, to assign responsibility to bystanders for the actions and inactions that support the commission of sexual assault, bystanders must be pictured and heard. They must be made visible and audible, such that their actions, and inactions, can be located onto identifiable bystander subjects who can be made accountable for it. At the same time, bystander intervention into sexual assault is also being made visible and audible as a counter strategy to the failures of other bystanders to intervene and arrest the enactments of sexual violence in which they participate. And as I have demonstrated, making bystanders visible and audible as responsible agents is enacted as a necessary part of the process, but it is ultimately an insufficient framework.

In the high-profile court case in which Stanford University undergraduate student Brock Turner was brought up on sexual assault charges for assaulting a highly inebriated woman, Chanel Miller, outside a campus party while she was in and out of consciousness, two graduate students from Sweden, Carl-Frederick Arndt and Peter Jonsson, intervened to stop the assault while they were biking on campus. In interviews, the men are reticent to take up the label of heroic interveners, humbly performing the responsibility they took as bystanders and encouraging others to read the powerful victim impact statement Chanel Miller read aloud to the court in 2016. In the *Washington Post* and in other press coverage, their actions are posed against coverage of Turner and the statements that character witnesses (members of his family and childhood friends) made to the court that claimed Turner was unfairly being singled out for responsibility for his rape of Miller. In his defense, Turner suggested that Miller, while unconscious, was able to consent to sexual contact, performing a particular model of sexually entitled White male irresponsibility for sexual assault committed against another person that blames the victim. Unlike many sexual assault cases, members of the public were able to see the bystanders who intervened to stop Turner's assault. News interviews and photographs that circulated in the press identify them as interventionist agents. In a *Washington Post* video posted online, rather than talk about himself or what happened, Jonsson encourages viewers to "spare a few minutes and read the letter written by the victim. To me it is unique in its form and comes as close as you possibly can get to putting words on the experience that words cannot describe."[98]

In her 2019 memoir *Know My Name*, Miller calls for a memorial to her survival and the bystanders who intervened to stop the assault, right at the site of her attack on campus:

> I encourage you to sit in that garden, but when you do, close your eyes, and I'll tell you about the real garden, the sacred place. Ninety feet away from where you sit there is a spot, where Brock's knees hit the dirt, where the Swedes tackled him to the ground, yelling, *What the fuck are you doing? Do you think this is okay?* Put their words on a plaque. Mark that spot, because in my mind I've erected a monument. The place to be remembered is not where I was assaulted, but where he fell, where I was saved, where two men declared stop, no more, not here, not now, not ever.[99]

While the cases analyzed in this chapter examine some of the ways bystander participation in sexual violence is being audio-visually recorded, Chanel Miller's writing and art model another framework for responding to sexual violence less focused on depiction, and more on the act of violence and responses to it.[100] She represents her own survival in relationship to the bystanders who intervened, and to other survivors who offered her support in the form of letters and social media posts, demonstrating forms of survivor-based justice that enact—and recognize the significance of—bystander responsibility.

In the Lonina and Parsons cases analyzed here, there is particular work being done to reveal bystander performances of irresponsibility in videos, livestreams, and digital photos. This is work many bystanders also do when they make and share digital mementos of their presence at and participation in gender violence; and it is work activists, legal experts, community groups, and survivors do in counter-narrating the meaning of violence that is recorded and distributed on bystander videos. When circulated and discussed, those videos and digital photos perform digitally recorded bystander irresponsibility as part of a social process, audio-visually establishing bystander responsibility for sexual violence, specifically sexual violence that is recorded. In this context of interpretation, the very notions of a duty to respond, report, and/or rescue another—a *duty to enact response-ability*—is based in a process of transforming bystanders who look at and document acts of violence into witnesses who can, and should, testify to them, whether in courts of law or in informal settings through online reporting, blogging, or other forms of social media participation.

Digital video and photography hold out a kind of promise that the perpetrators and witnesses to sexual violence will be held legally accountable for their roles in the commission of sexual assault because digital records help make their participation visible. And yet within current criminal justice systems, visibility cannot fulfill that promise alone. Across the cases discussed here, it was difficult to legally prove that sexual violence had occurred even when there was so much digital evidence to support it, a situation legal scholar Alexa Dodge suggests "further confirm[s] the criminal justice system's inability to provide something that feels like 'justice' for the vast majority of victims of sexual violence."[101]

This chapter has examined how the interaction between digital recordings, their circulation in news and social media and some of the commentary therein, and lawmaking provide different ways of understanding—and apprehending—bystander responsibilities for sexual violence. Bystanders matter in terms of how we address sexual violence because they are socially and physically proximate to violence, often as peers, friends, and acquaintances. Their position can be a fluid and changing one, sometimes working on behalf of, with, or in support of others who are directly assaulting someone; sometimes they become physical assailants themselves; and sometimes they act on behalf of the victim(s). Bystanders' proximity defines the limits and possibilities for their intervention, and their responsibility for sexual violence.[102] We may not always *see* the bystander in the media they create, yet bystander videos and photos nonetheless make legible what bystanders may have seen and heard via their recordings. Their videos and photos offer the possibility of digital witness, even if they alone cannot fulfill their potential to make bystanders responsible.

Notes

In addition to the editors of this collection, I would like to give special thanks to two of my interlocutors, Lisa Nakamura and Jonathan Sterne; Chahinez Bensari and Benjamin Nothwehr, for their for her crucial research assistance; and the audiences at the Law and the Visible speaker series at Amherst College in March 2018 and the annual meeting of the American Studies Association in November 2018. Research for this chapter was supported by the McGill University William Dawson Scholar fund.

1. Alexa Dodge, "The Digital Witness: The Role of Digital Evidence in Criminal Justice Responses to Sexual Assault," *Feminist Theory* 19, no. 3 (2018): 305. Dodge cites work from GraceAnn Carimico, Thuy Huynh, and Shallyn Wells, "Rape and Sexual Assault," *Georgetown Journal of Gender and the Law* 17, no. 1 (2016): 359–410.

2. Nicola Henry and Anastasia Powell, "Gender, Shame, and Technology-Facilitated Sexual Violence," *Violence against Women* 21, no. 6 (2015): 759.

3. See Tom Ley, "'She Is So Raped Right Now': Partygoer Jokes about the Steubenville Accuser the Night of the Alleged Rape," *Deadspin.com*, January 2, 2013, https://deadspin .com/she-is-so-raped-right-now-partygoer-jokes-about-the-5972527; Mairin Prentiss, "Small Community Names Names in Rehtaeh Parsons Case," *Global News*, April 11, 2013, https://globalnews.ca/news/474530/small-community-names-names-in-rehtaeh-parsons-case/.

4. Kelly Oliver, "Rape as Spectator Sport and Creepshot Entertainment: Social Media and the Valorization of the Lack of Consent," *American Studies Journal* 61 (2016): 2.

5. Judith Butler, *Giving an Account of Oneself* (New York: Fordham University Press, 2005), 85.

6. Kelly Oliver, *Witnessing: Beyond Recognition* (Minneapolis: University of Minnesota Press, 2001).

7. Carrie Rentschler, "Technologies of Bystanding: Learning to See Like a Bystander," in *Shaping Inquiry in Culture, Communication and Media Studies*, ed. Sharrona Pearl (London: Routledge, 2015), 15–40, quotation at 21.

8. See Kimberly K. Allen, "Guilt by (More Than) Association: The Case for Spectator Liability in Gang Rapes," *Georgetown Law Journal* 99 (2011): 837–67; Alexa Dodge, "Digitizing Rape Culture: Online Sexual Violence and the Power of the Digital Photograph," *Crime, Media & Culture* 12, no. 1 (2016): 65–82; Cressida J. Heyes, "Dead to the World: Rape, Unconsciousness, and Social Media," *Signs: Journal of*

Women in Culture and Society 41, no. 2 (2016): 361–83; Kelly Oliver, "Rape as Spectator Sport," 1–22.

9. See Henry and Powell, "Gender, Shame, and Technology-Facilitated Sexual Violence."

10. Molly Redden, "'It's Victimization': Push Grows to Charge Onlookers Who Tape Sexual Assaults," *The Guardian*, August 15, 2016, https://www.theguardian.com/society/2016 /aug/15/rape-prosecutions-onlookers-tape-sexual-assaults-legal-questions.

11. Holly Jeanine Boux and Courtenay W. Daum, "At the Intersection of Social Media and Rape Culture: How Facebook Postings, Texting and Other Personal Communications Challenge the Real Rape Myth in the Criminal Justice System," *University of Illinois Journal of Law, Technology & Policy*, no. 1 (2015): 149–86.

12. See Heyes, "Dead to the World"; Oliver, "Rape as Spectator Sport"; Susan Sontag, "Regarding the Torture of Others," *New York Times*, May 23, 2004; Judith Butler, "Torture and the Ethics of Photography," *Environment and Planning D: Society and Space* 25, no. 6 (2007): 951–66.

13. See Rebecca Adelman, *Figuring Violence: Affective Investments in Perpetual War* (New York: Fordham University Press, 2019).

14. Allen, "Guilt by (More Than) Association"; Heyes, "Dead to the World"; Oliver, "Rape as Spectator Sport."

15. See Carrie Rentschler, "Rape Culture and the Feminist Politics of Social Media," *Girlhood Studies* 7, no. 1 (2014): 65–82; Rentschler, "Technologies of Bystanding," 2015; Carrie Rentschler, "Bystander Intervention, Social Media Testimony and the Anti-Carceral Politics of Care," *Feminist Media Studies*, special issue "Affective Encounters," 17, no. 4 (2017): 565–84.

16. Maeve Duggan, "Online Harassment," *Pew Research Center*, Washington, DC, October 22, 2014; "Online Harassment 2017," *Pew Research Center*, Washington, DC, July 11, 2017.

17. See Liz Kelly, *Surviving Sexual Violence* (Minneapolis: University of Minnesota Press, 1988).

18. Jennifer Temkin and Barbara Krahé, *Sexual Assault and the Justice Gap: A Question of Attitude* (Portland, OR: Hart Publishing, 2008).

19. Elaine Craig, *Putting Trials on Trial: Sexual Assault and the Failure of the Legal Profession* (Montreal: McGill-Queen's University Press, 2018), 3.

20. Temkin and Krahé, *Sexual Assault and the Justice Gap*; Craig, *Putting Trials on Trial*; Lise Gotell, "Reassessing the Place of Criminal Law Reform in the Struggle against Sexual Violence: A Critique of the Critique of Carceral Feminism," in *Rape Justice: Beyond the Criminal Law*, ed. Anastasia Powell, Nicola Henry, and Asher Flynn (London: Palgrave Macmillan, 2015), 53–71.

21. Temkin and Krahé, *Sexual Assault and the Justice Gap*, 2.

22. See Donna Coker, "Transformative Justice: Anti-Subordination Processes in Cases of Domestic Violence," in *Restorative Justice and Family Violence*, ed. Heather Strang and John Braithwaite (Cambridge: Cambridge University Press, 2002), 128–52; Angela Davis, *Are Prisons Obsolete?* (Boston: South End Books, 2003); Angela Harris, "Heteropatriarchy Kills: Challenging Gender Violence in a Prison Nation," *Washington University Journal of Law & Policy* 37 (2011): 11–65; "Incite! Women of Color against Violence," *Statement on Gender Violence and the Prison Industrial Complex* (Redmond, WA: Community Accountability Working Document: Principles /Concerns/Strategies/Models, 2003), https://www.nasco.coop/sites/default/files /srl/InciteWomenofcoloragainstviolence.pdf; Mimi Kim, "Moving Beyond Critique: Creative Interventions and Reconstructions of Community Accountability," *Social Justice* 37, no. 4 (2010): 14–35; Beth Richie, *Arrested Justice: Black Women, Violence, and*

America's Prison Nation (New York: New York University Press, 2012); Beth Richie, "Law Enforcement Violence against Women of Color," in *Color of Violence: The Incite! Anthology*, edited by Incite! Women of Color against Violence (Durham, NC: Duke University Press, 2016), 138–56; Julia Sudbury, ed., *Global Lockdown: Race, Gender and the Prison-Industrial Complex* (New York: Routledge, 2005).

23. See Jeremy Waldron, "On the Road: Good Samaritans and Compelling Duties," *Santa Clara Law Review* 40, no. 4 (2000): 1053–1103.

24. John Kaplan, "A Legal Look at Prosocial Behavior: What Can Happen for Failing to Help or Trying to Help Someone," *Journal of Social Issues* 28, no. 3 (1972): 219–20. Most law-making regarding the offering of aid to others does not establish duties to assist but rather offers protection to those who do offer assistance from civil liability for any harm they may cause while attempting to help someone in an emergency. In the context of Anglo-American law, "the failure to act provides a ground for criminal sanctions only when there is a pre-existing legal duty to act," such as those tied to one's professional duties and obligations. Outside of the bounds of these roles, failures to act are not considered "conduct" in a legal sense. This is unlike many European legal systems, which require "bystanders to render easy aid to those gravely imperiled, provided he [*sic*] has recognized the opportunity to give assistance and has the ability to do so." See John Kleinig, "Criminal Liability for Failures to Act," *Law and Contemporary Problems* 49, no. 3 (1986): 161–80.

25. Kathleen Ridolfi, "Law, Ethics, and the Good Samaritan: Should There Be a Duty to Rescue?" *Santa Clara Law Review* 40, no. 4 (2000): 958.

26. See Barbie Zelizer, *Remembering to Forget: Holocaust Memory through the Camera's Eye* (Chicago: University of Chicago Press, 1998).

27. See Powell, Henry, and Flynn, eds., *Rape Justice: Beyond the Criminal Law*; Anastasia Powell, "Seeking Informal Justice Online: Vigilanteism, Activism, and Resisting a Rape Culture in Cyberspace," in *Rape Justice: Beyond the Criminal Law*, 218–37; Rentschler, "Bystander Intervention, Social Media Testimony and the Anti-Carceral Politics of Care"; Michael Salter, "Justice and Revenge in Online Counter-Publics: Emerging Responses to Sexual Violence in the Age of Social Media," *Crime, Media & Culture* 9, no. 3 (2013): 225–42.

28. Greg Stratton, Anastasia Powell, and Robyn Cameron, "Crime and Justice in Digital Society: Towards a 'Digital Criminology,'" *International Journal for Crime, Justice and Social Democracy* 6, no. 2 (2016): 24.

29. See Scott Veitch, *Law and Irresponsibility: On the Legitimation of Human Suffering* (London: Routledge, 2008), 41.

30. Martha Merrill Umphrey, "Law in Drag: Trials and Legal Performativity," *Columbia Journal of Gender and Law* 21, no. 2 (2012): 114–29.

31. Nicola Gavey, *Just Sex? The Cultural Scaffolding of Rape* (London: Routledge, 2004); Lynn Bowes-Sperry and Anne O'Leary-Kelly, "To Act or Not to Act: The Dilemma Faced by Sexual Harassment Observers," *Academy of Management Review* 30, no. 2 (2005): 288–306; V. L. Banyard, "Who Will Help Prevent Sexual Violence: Creating an Ecological Model of Bystander Intervention," *Psychology of Violence* 1, no. 3 (2011): 216–29.

32. Robert Abele, "Review: 'Roll Red Roll' Examines the Steubenville Rape Case and a Culture of Toxic Masculinity," *Los Angeles Times*, April 3, 2019 (emphasis added).

33. See Ley, "'She Is So Raped Right Now.'" The entire video can still be viewed on the *Deadspin* website.

34. Oliver, "Rape as Spectator Sport," 4.

35. See Whitney Phillips, "It Wasn't Just the Trolls: Early Internet Culture, 'Fun,' and the Fires of Early Internet Culture," *Social Media + Society* (2019): 1–4.

36. Image reproduced in Max Read, "Instagram, YouTube-Fueled High School Rape Trial Begins Today," *Gawker*, March 13, 2013.
37. See Ley, "'She Is So Raped Right Now,'" and Oliver, "Rape as Spectator Sport."
38. Judith Butler, *Precarious Life: The Powers of Mourning and Violence* (New York: Verso, 2004).
39. See Read, "Instagram, YouTube-Fueled High School Rape Trial Begins Today."
40. Oliver, "Rape as Spectator Sport."
41. Jane Anderson and Supriya Prasad, "Prosecuting Image Exploitation," *Aequitas Strategies: The Prosecutor's Newsletter on Violence against Women* 15 (2015): 3.
42. Anderson and Prasad, 3.
43. Connor Simpson, "The Kids at the Steubenville Rape Party Told Cops They Should Have Stopped It," *The Atlantic*, March 24, 2013.
44. See Alexander Abad-Santos, "Who's in Trouble Next in Steubenville?" *The Atlantic*, March 18, 2013; Simpson, "The Kids at the Steubenville Rape Party."
45. Anderson and Prasad, "Prosecuting Image Exploitation," 5.
46. Renu Mandhane, "Duty to Rescue through the Lens of Multiple-Party Sexual Assault," *Dalhousie Journal of Legal Studies* 9 (2000): 2.
47. Mandhane, "Duty to Rescue," 11.
48. Redden, "'It's Victimization.'"
49. Anderson and Prasad, "Prosecuting Image Exploitation," 3; Arielle Pardes, "#JadaPose is about More Than Just Social Media," *Refinery 29*, July 16, 2014, https://www.refinery29.com/2014/07/71205/jadapose-sexual-assault-social-media. I use the term "victim" to refer to the position of having experienced violence and its often-traumatizing effects as the direct target of it. I use "survivor" to name the positionality of those who reclaim their victimization experience as part of a process of healing and survival. Both are forms of identification that have been and can be politicized, as we see in the case of victims' rights movements and feminist survivor-centered movements against gender-based violence, respectively. In some of the cases of sexual assault and online image exploitation discussed in this chapter, the victims did not, in the end, survive (e.g., Amanda Todd, Rehtaeh Parsons); therefore, I refer to them using the term "victim" when not referring to their proper names. See Carrie Rentschler, *Second Wounds: Victims' Rights and the Media in the U.S.* (Durham, NC: Duke University Press, 2011).
50. Heyes, "Dead to the World," 372.
51. Boux and Daum, "At the Intersection of Social Media and Rape Culture."
52. Oliver, "Rape as Spectator Sport."
53. Allen, "Guilt by (More Than) Association," 839.
54. Mike McPhate, "Teenager Is Accused of Live-Streaming a Friend's Rape on Periscope," *New York Times*, April 18, 2016.
55. Associated Press, "Ohio Teen Sentenced to 9 Months in Prison in Livestream Rape Case," *CBC*, February 14, 2017, http://www.cbc.ca/news/world/periscope-livestream-assault-plea-1.3983377.
56. Redden, "'It's Victimization.'"
57. See Lauren Messman, "The Teen Who Livestreamed Her Friend's Rape Got Nine Months in Prison," *Vice*, February 14, 2017, https://www.vice.com/en_us/article/qkxpaq/the-teen-who-livestreamed-her-friends-rape-got-nine-months-in-prison.
58. Charlotte England, "Teenager Jailed for Broadcast of Girl's Rape on Online Periscope App," *Independent* (UK), February 15, 2017, https://www.independent.co.uk/news/world/americas/teenager-marina-lonina-livestream-rape-17-year-old-friend-periscope-app-sentence-prison-columbus-a7581196.html; Messman, "The Teen Who Livestreamed."

59. See Sarah Banet-Weiser, *Authentic TM: The Politics and Ambivalence in Brand Culture* (New York: New York University Press, 2012).

60. See Alice Marwick, *Status Update: Celebrity, Publicity and Branding in the Social Media Age* (New Haven, CT: Yale University Press, 2013); Lee Humphreys, *The Qualified Self: Social Media and the Accounting of Everyday Life* (Cambridge, MA: MIT Press, 2018).

61. National Public Radio, "Live Streaming of Alleged Rape Shows Challenges of Flagging Video in Real Time," WLRH, Huntsville, AL, April 19, 2016, http://www.wlrh.org/ NPR-News/live-streaming-alleged-rape-shows-challenges-flagging-video-real-time.

62. National Public Radio, "Live Streaming of Alleged Rape."

63. Nora Caplan-Bricker, "Why Does the Teen Who Live-Streamed an Attack on a Friend Face the Same Charges as the Alleged Rapist?" *Slate.com*, April 19, 2016, https:// slate.com/human-interest/2016/04/is-the-teen-who-live-streamed-an-alleged-rape -an-accomplice.html.

64. Lucy Clarke-Billings, "Depraved Teenage Girl Who Filmed and Livestreamed Her Friend Being Raped Is Jailed for Just 9 Months," *Mirror* (UK), February 16, 2017.

65. Andrew Welsh-Huggins, "Ohio Teen Pleads Not Guilty to Livestreaming Friend's Rape," *NBC TV Chicago*, April 19, 2016.

66. Megan Sapnar Ankerson, "Periscope: The Periscopic Regime of Live-Streaming," in *Appified: Culture in the Age of Apps*, ed. Jeremy Wade Morris and Sarah Murray (Ann Arbor: University of Michigan Press, 2018), 228.

67. See Maura Conway and Joseph Dillon, "Case Study: Future Trends: Live-Streaming Terrorist Attacks?" *VoxPol* (Dublin, Ireland: Network of Excellence for Research in Violent Online Political Extremism, 2016).

68. See Conway and Dillon, "Case Study: Future Trends," 4.

69. Ankerson, "Periscope," 228.

70. Humphreys, *The Qualified Self*, 5.

71. Humphreys, 5.

72. Paul Frosh, "The Gestural Image: The Selfie, Photography Theory and Kinesthetic Sociability," *International Journal of Communication* 9 (2015): 1610.

73. See Alice Marwick, *Status Update*; Zeynep Tufecki, "'Not This One': Social Movements, the Attention Economy, and Microcelebrity Networked Activism," *American Behavioral Scientist* 57, no. 7 (2013): 848–70; Stuart Allan, *Citizen Witnessing: Revisioning Journalism in Times of Crisis* (Cambridge: Polity Press, 2013).

74. Messman, "The Teen Who Livestreamed."

75. Allie Conti, "Does Livestreaming a Rape Make You a Rapist?" *Vice*, April 20, 2016, https:// www.vice.com/en_us/article/dp5mmz/does-live-streaming-a-rape-make-you-a-rapist.

76. Conti, "Does Livestreaming a Rape Make You a Rapist?"

77. See "Marina Lonina Pleads Not Guilty," in Mike McPhate, "Teenager Is Accused of Live-Streaming a Friend's Rape on Periscope," *New York Times*, April 18, 2016.

78. Sharon Sliwinski, *Human Rights in Camera* (Chicago: University of Chicago Press, 2011), 4.

79. The Youth Criminal Justice Act prevents the press from making the names of the young men public because they were minors. Normally the victim's name would be banned too, but Parsons's parents won an exception to the law so that they could use their daughter's name while talking about her case. Rehtaeh Parsons's name was under a publication ban until December 17, 2014, when the Prosecutor's Office in Nova Scotia agreed to her parents' requests to be able to use her name in speaking to the press and in other public venues about their daughter after her death.

80. Melanie Patten, "Child Pornography Charges Laid in Rehtaeh Parsons Case," *Global News* and *Canadian Press*, August 8, 2013, https://globalnews.ca/news/770393 /child-pornography-charges-laid-in-rehtaeh-parsons-investigation/.

81. Christie Blatchford, "Boy in Notorious Rehtaeh Parsons Case Talks for the First Time about What Happened," *National Post* (Canada), January 15, 2015, https://nationalpost .com/opinion/christie-blatchford-boy-in-notorious-rehtaeh-parsons-photo-talks -for-first-time-about-what-happened; After Parsons's death, an online dating company, ionechat.com, used a profile photo of her to advertise its services on Facebook with ad copy that stated "Find Love in Canada!"—constituting another image-based violation of Parsons and her family. See Andre Mayer, "Rehtaeh Parsons Facebook Ad a Textbook Case of Online Photo Abuse," *CBC News*, September 19, 2013, http://www.cbc.ca /news/rehtaeh-parsons-facebook-ad-a-textbook-case-of-online-photo-abuse-1 .1859585.

82. Charlie Gillis, "Rehtaeh Parsons and the Problem of Bystanders," *Maclean's*, April 18, 2013, http://www.macleans.ca/news/canada/a-deafening-silence/.

83. Wendy Gillis, "Rehtaeh Parsons: A Family's Tragedy and a Town's Shame," *Toronto Star*, April 12, 2013, https://www.thestar.com/news/canada/2013/04/12/rehtaeh _parsons_a_familys_tragedy_and_a_towns_shame.html. At this date, it is difficult to locate the photo through a Google image search, which suggests that it has been scrubbed from the Internet and delinked from search terms using her name. Under the Intimate Images and Cyber Protection Act of Nova Scotia that was passed in 2018, complainants can request the removal of intimate images that have been posted online in order to harm them. The province established a Cyber Scan unit to conduct image removals of photos circulated online for cyber harassment purposes. Manitoba and Alberta also have similar laws on the books. For more on this, see Michael Tutton, "New Cyberbullying Law Can Force Removal of Intimate Images Online," *CBC.ca*, July 5, 2018. https://www.cbc.ca/news/canada/nova-scotia /nova-scotia-new-cyberbullying-law-investigators-1.4736021.

84. Blatchford, "Boy in Notorious Rehtaeh Parsons Case Talks."

85. Blatchford.

86. See Dodge, "Digitizing Rape Culture."

87. Oliver, "Rape as Spectator Sport."

88. Blatchford, "Boy in Notorious Rehtaeh Parsons Case Talks"; See Patten, "Child Pornography Charges."

89. Christie Blatchford, "New Details in Rehtaeh Parsons Case Show Why Police Didn't Lay Charges," *Canada.com*, April 26, 2013, http://o.canada.com/news/national/blatchford -even-the-rehtaeh-parsons-case-has-more-than-one-side (emphasis added).

90. Blatchford, "New Details in Rehtaeh Parsons Case."

91. Dodge, "The Digital Witness," 304.

92. Julia Nicol and Dominique Valiquet, *Legislative Summary of Bill C13: An Act to Amend the Criminal Code, the Evidence Act, the Competition Act, and the Mutual Legal Assistance in Criminal Matters Act*, August 28, 2014 (Ottawa, CA: Library of Parliament, 2014); Rachel Brydolf-Horwitz, "Embodied and Entangled: Slow Violence and Harm via Digital Technologies," *Environment and Planning C: Politics and Space* (2018): 1–18; A. Wayne McKay, "Law as an Ally or Enemy in the War on Cyberbullying: Exploring the Contested Terrain of Privacy and Other Legal Concepts in the Age of Technology and Social Media," *University of New Brunswick Law Journal* 66, no. 3 (2015): 1–51. The original 2013 Nova Scotia "Cyber Safety Act" was struck down after a constitutional challenge in 2015; the 2018 "Intimate Images and Cyber-protection Act" replaced it. See Brett Ruskin, "Court Strikes Down Anti-Cyberbullying Law Created after Rehtaeh Parsons's Death," *CBC.ca*, December 11, 2015, https://www.cbc.ca/news /canada/nova-scotia/cyberbullying-law-struck-down-1.3360612; see also Nova Scotia, "Intimate Images and Cyber-protection Legislation Proclaimed," July 5, 2018, https:// novascotia.ca/news/release/?id=20180705004.

93. See Jane Bailey, "A Perfect Storm: How the Online Environment, Social Norms, and Law Shape Girls' Lives," in *eGirls, eCitizens: Putting Technology, Theory and Policy into Dialogue with Girls' and Young Women's Voices,* ed. Jane Bailey and Valerie Steeves (Ottawa, ON: University of Ottawa Press, 2015), 21–54.

94. See Paul Frosh and Amit Pinchevski, "Introduction: Why Media Witnessing? Why Now?," in *Media Witnessing: Testimony in the Age of Mass Communication,* ed. Paul Frosh and Amit Pinchevski (Basingstoke: Palgrave Macmillan, 2009), 1–19.

95. Rentschler, "Technologies of Bystanding," 17–18.

96. Veitch, *Law and Irresponsibility.*

97. See Umphrey, "Law in Drag."

98. See video by Monica Akhtar published in Lindsay Bever, "The Swedish Stanford Students Who Rescued an Unconscious Sexual Assault Victim Speak Out," *Washington Post,* June 8, 2016.

99. Chanel Miller, *Know My Name: A Memoir* (New York: Viking, 2019), 313.

100. Chanel Miller, "I Am With You" (video), September 24, 2019, https://youtu.be/ouIxvBMF7Rw; see also Miller, *Know My Name.*

101. Dodge, "The Digital Witness," 318.

102. Rentschler, "Technologies of Bystanding," 21.

CHAPTER 4

Between the Bodycam and the Black Body

The Post-Panoptic Racial Interface

EDEN OSUCHA

> The technologies of image-capturing, its instruments from *camera obscura* to smartphones and their products—prints, film, jpegs—were not designed for Black life. And they are obsessed with Black death.
>
> —Nehal El-Hadi, "Death Undone"

> because white men can't
> police their imagination
> black men are dying
> —Claudia Rankine, Rankine, ["because white men can't"],
> *Citizen: An American Lyric*

Over the last decade, police departments have increasingly adopted body-worn cameras—colloquially, "bodycams"—as standard-issue equipment, in addition to video cameras mounted on patrol-car dashboards. On the other side of police-civilian encounters, an informal net of sousveillance has proliferated through smartphone cameras and the Internet, practices of near-instantaneous uploading and even livestreaming of video via social media platforms.[1] The images and video produced by these visual technologies have played a constitutive and transformative role in broader public conversations about the American criminal justice system's racial disproportionalities. Bystander cellphone footage, in particular, has galvanized broad public protest against racially motivated police violence under the aegis of the now-international Movement for Black Lives. Until quite recently, police body camera programs were central among the demands of protest organizers; they still predominate legislative efforts at police

93

reform and are broadly popular with the general public. Meanwhile, video recordings of the deaths of unarmed African Americans at the hands of police have continued to accumulate in the digital public sphere. Filming violent policing, it seems, does little to deter it.

It's hard to conclude otherwise after viewing video footage of the police torture and killing of Black Minneapolis resident George Floyd on May 25, 2020. In full view of his three colleagues' body cameras, the cellphones of outraged civilian witnesses, and security cameras, Derek Chauvin kneeled on Floyd's neck for eight minutes and forty-six seconds, during which Floyd pleaded for his life, called out for his dead mother, and slowly asphyxiated. In its profusion of viewpoints, unprecedented intimacy, and sheer temporal excessiveness, the composite portrait of the killing provided by these videos radically shifted the conversation around race and policing in the United States. The national protests following Floyd's killing called for defunding rather than reforming police departments. But abolitionist agendas have little traction in national policy discussions. Proposals to equip officers with bodycams, however, have increased at every level of government. Legislation funding such programs is, as of this writing, the only federal policing reform measure likely to pass, having broad bipartisan support.

The bodycam's prominent role in discussions of police violence reveal the intersections of law, visuality, and race that form the discursivity of racial blackness in the United States. This is a racial formation that derives from slavery, as I discuss, and is therefore fundamentally panoptic. How does the bodycam—a technology that at once collates legal, visual, and racial epistemologies—fit into or extend this history? The answer is in slavery's visual cultures. These encompass the institutionalized forms of the auction block, plantation oversight, and the ritualized performance of slave subjection—and post-emancipation popular culture's burlesquing of enslavement and mastery in minstrel theater and the racial grotesquerie found in late nineteenth-century commodity forms and commercial imagery. Even outside the architectonic disciplinarity of the plantation or enslaving domestic spaces, the lines of enclosure that defined the panoptic configuration of "race" as a visual social episteme during slavery were unmistakable. In the aftermath of slavery in the United States, these structures surfaced in the enforcing of racial hierarchies through laws that reproduced racial visuality through the racialization of social space via vagrancy and loitering laws, segregation in public institutions and commercial establishments, and real estate red-lining.

The reproduction of racial hierarchies through the segregation of space and personal mobility in each of these instances indicates the operations of power characteristic of the "disciplinary societies" identified by Michel Foucault. Gilles Deleuze uses this phrase to summarize Foucault's historiographic project in which modern social institutions of the school and the factory are understood in terms of their carceral features—namely, the dynamics of enclosure and fixed organization of power for which the Benthamite prison model, the Panopticon, is exemplary. As with the prison in which individual cells are perpetually available to surveillance by an omniscient, centrally located guard tower, surveillance becomes, in the institutions that compose disciplinary societies, internalized by individuals in relation to clear lines of power. In the United States, these institutions also serve as sites for the regulation and production of racial hierarchies. Schools, the workplace, and prisons, especially, are central sites for the reproduction of racialized asymmetries of power, privilege, and precarity.

I am arguing that the bodycam is an emblematic technique of power for what Deleuze formulates as the new "societies of control."[2] Indeed, it is paradigmatic for the organization and operation of power in the present, just as Jeremy Bentham's prison architecture blueprint is for Michel Foucault's panoptic theory of society. In post-panoptic societies, power is concentrated at shift points that are multimodal and highly mutable. Asymmetries of power, wealth, and access, and thus targeted vulnerabilities, are reproduced in terms that are always shifting and difficult to locate as fixed "systems." Power, Deleuze notes, is no longer emblematized by fixed structures or their physicalities but by "mutable, contingent effects."

The bodycam exemplifies this new system of domination. This is not simply because the bodycam, as a surveillance technology, literally has no fixed address. It is because the bodycam is discursively constructed as having a singular, transparent, fact-finding gaze even though the images it produces are subject to myriad contingencies: whether the camera is turned on, whether footage has been converted to data and stored and backlogged, and where the officers are standing or moving in relation to the citizens with whom they interact. And then there are policies that differently shape the treatment of the data and regulate or restrict access to the images or determine how they are used as evidence. In contrast to the mutability and immateriality of state power and the bodycam's distributed, multi-modal operations of power, the Black bodies that become this technology's emblematic targets are fixed objects, framed before they even come into the camera's view. Their positioning in these terms is integral to the discursive construction of the bodycam and its visuality. This is clear in the post-Ferguson rhetorics around police brutality and misconduct, police reform, and police professionalism.

The bodycam's gaze is wholly machinic rather than directed by the "artistic" whims of the user. It is a camera without a photographer. As such, it does not just capture what the wearer sees, it documents the very perspective of the officer. It does so seemingly without mediation, for it is not positioned between the camera operator's eye and the object being filmed. It stands in for the eye of the officer, even though the camera's point of view is well outside its user's actual sightlines; nevertheless, it is regarded as synonymous with them—the footage that it captures, synonymous with what the officer wearing it saw. As I discuss below, this is enormously significant for how bodycam footage is used as evidence in court and how it relates to policy testimony.

The illusion of interchangeability between the bodycam's and officer's points of view is secured through how bodycam footage seems to reproduce the experience of live, unmediated perspective: the camera movement is kinetic, given the movement of the wearer in space, including alterations of gait, speed, and position; the audio recording of the voice of the wearer is more proximate than other voices; the user's body is visible in fragments, mostly as arms and hands. In short, bodycam footage registers the embodied presence of the user's subjectivity. By foregrounding the officer's subjectivity as the locus of visual perception, the bodycam produces a representational field in which the civilian-subjects with whom the officer interacts are perpetually objectified, available for intimate surveilling by the state whose authoritative gaze is at once reified by the camera and dissimulated.

Unlike other technologies of surveillance, the bodycam does not extend or augment human visual perception. It would be more precise to say that the bodycam's intended function is as an evidence-gathering device. Law has long regarded camera-based media and photographic objectivity in terms of an epistemology of naive realism.[3] For example, there is in common law a distinction between the probative value of videos, films, and photographs and that of pictorial representations made by "hand." The former, notes Ira Torresei, are admissible in court as "real evidence," while someone's sketch of a crime scene—no matter how detailed or verisimilar—would be treated as a variety of hearsay, akin to an oral statement made outside a trial or legal hearing.[4] Like a photograph, legal testimony is the product of mediation—in the form of the rules and penalties by which what would otherwise be regarded as a kind of *self*-narration becomes (under oath and penalty of perjury) depersonalized and thus evidential.[5]

As a genre of legal evidence, the camera-made image's impersonality is presumed. Photographs require no swearing in. Their affidavits are immune to charges of perjury. What Walter Benjamin famously termed "mechanical reproduction" generates an image the court may regard as indelibly "true" by sheer virtue of its technological origins, which ostensibly decouples the material object-image from human sight.[6] The assumption that technologized vision is inherently objective is implied in the categorical treatment of photographic and video images as "real evidence." This designation is otherwise reserved for material objects that corroborate or disprove an argument about the facts of a case and thus, in its applications to camera-made images, denotes a tangible connection between a picture and what it depicts. In court, the photograph (or the digital video produced by a body-worn camera) serves as a generic sign of "facticity" as well as a specific sign of the facts of a given case. In this, it resembles the fingerprint that serves as a sign of the finger having actually been "there" to make the print and thus, as evidence, is regarded as inextricable from the particular finger that made it.

That the camera-made image continues to be privileged as a genre of legal evidence in the era of digital media is a relic of the early history of photography and what emerged as the medium's distinctive social logics. Originally, the medium produced images that bore the trace of the material presence of that which they depicted, its specific location in time and place: the camera's shutter opening at a particular moment in time, this particular object's position before the lens in that particular place. In this way, the photograph was very much like a fingerprint. Considered as forms of representation, both the analog photograph and the fingerprint exemplified the characteristics of the "index"—in the sense of what nineteenth-century philosopher Charles Sanders Peirce meant by the term.

Writing in the era of photography's explosion in both popularity and in non-portrait, documentary uses, Peirce developed a theory of the "index" as a distinctive category of signification within his broader account of representation and meaning. A painting would fall into the category of sign Peirce deemed the "icon," in which a representation is linked to its referent by likeness. Unlike icons or "symbols"—in which the relation between representation and referent is a matter of convention (as with words and their meanings or traffic signs), the index, as understood by Peirce, describes a relationship between representation and referent that is causative. It directs our attention to a thing without describing it: "The index asserts nothing; it only says 'There!'"[7] In the digital era, the camera-made image is no longer technically dependent upon material presence for its composition. Yet, indexicality remains an essential quality of the medium as a social form, as reflected in the popularity of the digital "selfie" as global mass culture's favorite practice of self-elaboration.

Along with its indexical function as a sign of "having been there," the social meaning of photography that informs how videos are positioned as legal evidence also derives in earlier historical institutional and bureaucratic uses of photography. Toward the end of the nineteenth century, photography was commonly regarded as a source of medical and scientific knowledge. Correlatively, the medium played an instrumental role in the earliest attempt at what today is known as data-based policing. I refer, of course, to the invention of the mugshot, a practice first introduced in France in 1888.

The modern discipline of criminology emerged from the archives of police arrest photographs and sought in the faces of the accused a common social profile. In a previous essay on the historical relationship between early legal discourses of privacy in the United States and visual discourses of race, I describe how these adjacent disciplinary uses of photography, which traded upon what Alan Sekula terms the medium's "instrumental realism," served to systematically reify racist ideology, in particular, as objective social reality. It did so by way of the aura of indexicality at the heart of medium's unparalleled signifying capacity:[8] "Constituting the supposedly material evidence of [racial] difference, photographic archives, such as Francis Galton's infamous composite photographs of racial types or Alphonse Bertillon's biometrics practices [aimed at identifying 'criminal' physiognomies], simultaneously confirmed the suppositions of science, the state, and cultural 'common sense' and provided an external basis for those knowledges."[9] Photography did not generate nor did it refine the conceptual entanglements of race, sexuality, and disability underlying modern discourses of the criminal. Its purpose, in this historical moment, was to consolidate and reassemble categories of social difference into a visual grammar of "the criminal" for the purposes of public surveillance. This was, in effect, the invention of police practices of (racial) profiling.

This history haunts what Andrew Goldsmith terms "policing's new visibility."[10] Unlike the "old visibility" of scientific racism and mug shots—a representational regime produced by and located within official, institutional (state and academic) image-making practices—the transformations in the practices, cultures, and understanding of US policing to which Goldsmith alludes emerged within the ubiquitous techno-visibility of the present. Policing's new visibility is a product, in other words, not simply of the practices of video recording and surveillance used by police departments but of the interplay between these practices and the now routine use of a personal consumer technology—the smartphone—as a means of recording eyewitness documentation of officer use-of-force incidents.

Photography's status as a "realistic" medium informs how race itself is seen and how it has been produced as an organizing category of social and cultural meaning within modern epistemologies of the visual. Black bodies are made to constitute—in frequently violent and dehumanizing terms— the loci of race's visual epistemologies, which implicates photography and video's historically privileged positioning within discourses of the real. When Black bodies come into view in the photographic and video images called upon as evidence in court, the indexical function that is essential to these media discourses is redoubled by the racial indexicality—that is, the idea of race as visible, that likewise resides in the surface of their forms.[11] In such instances, the presumed self-evidence of photographic meaning dissimulates how the discursivity of race shapes the form of the image's "truth" before the law.

Although, speaking both practically and ideologically, I am arguing that bodycams are a surveillance technology, their widespread adoption in what policing scholars and activists deem the post-Ferguson era is due largely to claims about the bodycam's efficacy in promoting transparency and accountability in policing.[12] These claims frame the police body camera as a tool of counter-surveillance for which the civilian-wielded smartphone is actually, in practice, paradigmatic. As legal scholar Carrie Myers Morrison observes, "Body camera proponents contend that video recording . . . improves professionalism, provides material for training and evaluation, and serves as evidence when the police are accused of wrongdoing."[13] (These claims are central to how manufacturers market body-worn cameras to police departments.) At this writing, the national protest movement against anti-Black police violence, revitalized by widespread outrage over the viral footage of the murder of George Floyd recorded on bystander cellphone, has placed defunding and abolition of municipal police departments at the center of their demands. Prior to this moment, though, police reform activists, including those affiliated with the Movement for Black Lives, centered the bodycam as a tool for prosecuting and preventing police violence. At the federal level, the George Floyd Justice in Policing Act of 2020 recently passed by the House of Representatives (though poised to be voted down in the Senate or vetoed by the current president) includes a provision making bodycam programs mandatory, suggesting that the technology remains aligned in the broader political imagination with police accountability, transparency, and reform.

In practice, how video depictions of alleged police misconduct are presented as evidence for consideration by a grand or trial jury contradicts any assumptions about the transparency of what the bodycam "reveals." Juries are explicitly asked to treat these texts as *non*-self-evident and in need of interpretation—that is, they are instructed to mistrust their ability to make sense of what they see when they look at these images; by contrast, the public consumption of these same images—by activists, news media, and ordinary social media users—presupposes their self-evidence in the eyes of the civilian nonexpert. Thus, in the representation of video footage to a jury, the moving image is remediated as a series of still shots in frame-by-frame analyses. The singularity of a photographic image dissolves in the plaintiff's or defendant's counsel's performance of a legal hermeneutics that emphasizes the image's multiplicity, enlarging or cropping the image in order to configure its meaning around a single detail.

Law, like technologies of visuality, is both suffused with racial meaning and is itself racializing in its significations. Indeed, the social inscription of race might be read as one index of how law and the visual operate in mutually codetermining relation; the ocularity and corporeality of race informing how the Black body "appears" within the courtroom, just as, *mutatis mutandis*, legal concepts used to frame visual images as "evidence" effectively screen their meaning. The two surveillance systems in which the police body camera participates create what we might call the *post-panoptic* interface of US racialization, as in combination they mediate the encounter between police and the Black or Brown citizen and also the police video's construction and interpretation within the law. These two systems are, namely, the racialized surveillance of Black and Brown communities and individual bodies, which experience law enforcement and criminal justice oversight disproportionately, and the Internet, through which social media and video streaming sites install the images and videos of police violence within a "virtual" literalization of the system of photographic portraiture analyzed by Sekula in his classic essay "The Body and the Archive."

Writing in 1986, Sekula describes a dialectical relationship between the "honorific" protocols of individual photographic portraiture as they develop over the nineteenth century and a "repressive" instrumentalization of photography in policing and criminal science. They constitute a single system, "capable of functioning both honorifically and repressively."[14] Through the production and archiving of criminal portraits and visualized technologies for articulating knowledge about (frequently racialized) criminal "otherness," photographic portraiture performed what Sekula describes as "a role no painted portrait could have performed in the same thorough and rigorous fashion": "This role derived not from any honorific portrait tradition, but from the imperatives of medical anatomical illustration. Thus photography came to establish and delimit the terrain of the *other* . . . to the extent that the legal basis of the self lies in the model of property rights, in what has been termed 'possessive individualism,' every proper portrait has its lurking, objectifying inverse in the files of the police."[15] Photography in this way "introduce[d] the panoptic principle into daily life." The "honorific" and "repressive" functions were coterminous, constituting a "social and moral hierarchy" that framed the meaning of these seemingly diametric archives: "Every portrait implicitly took its place within a social and moral hierarchy. The private moment of sentimental individuation, the look at the frozen gaze-of the-loved-one, was shadowed by two other more public looks; a look up, at one's betters, and a look down at one's inferiors."[16]

Today, what Sekula terms "more public looks" are found not in the archives of the police but the constitutive modalities of spectatorship and self-display that characterize social media, discursive realms in which "sentimental individuation" transpires as *public*. Nineteenth-century photography's "panoptic principle" becomes, in the contemporary media environment of the Internet and its semantic field, something more properly described as *post-panoptic*. This distinction registers how, in digital culture, the honorific and the repressive functions that constitute the essential binary unit of photographic meaning described by Sekula can be simultaneously present in a single image. This function is exemplified by the digital afterlives of bodycam and other eyewitness video of fatal civilian-police encounters, images variously framed as activist tools, news items, and spectacle entertainment. These videos circulate as both honorific images meant to restore dignity, individuality, and value to the dead and as repressive images that inscribe their subjects within dehumanizing, depersonalizing archives of racialized criminality. One of the characteristics of this post-panoptic interface is the extreme mediagenicity of the images captured by bodycams, which transform these legal documents' disciplinary and regulatory function into modes of aesthetic consumption. If the "everyday" panopticism theorized by Sekula enfolds photography's repressive function into its honorific significations, in the post-panoptic discursive field, the repressive is wholly dissimulated within the aesthetic.

***

The social and political transformations that led to police work being reconceptualized in the American imagination in terms of technovisibility were well underway before the fatal shooting of Michael Brown on August 9, 2014, in Ferguson, Missouri. The relatively new ubiquity of camera-equipped smartphones, combined with the ability to easily upload images and video to wide-reaching social media platforms, contributed to an intensifying net of informal sousveillance of police, foregrounding eyewitness documentation of use-of-force incidents between officers and civilians. Likewise, increasing numbers of officers were by then equipped with body-worn cameras.

A significant factor in both increased implementation of and public awareness of police body cameras at the time of Brown's death was the steadily growing interest in bodycam technology on the part of US police and sheriff's departments. This interest was fed in no small part by preliminary findings from a widely heralded 2012 study that seemed to prove the cameras' effectiveness in deterring police brutality. The findings in the Rialto study, as it has since become known, promote the idea that by implementing body camera programs, departments could deter and thereby prevent police misconduct.[17] Central to this claim was the belief that bodycams could furnish prosecutors in police misconduct cases with credible evidence.

At the time of Brown's death, the manufacturer then still known as Taser International was in the process of realigning its brand identity with what is now the market-dominant police bodycam, technology it modeled on the weapon-mounted "Taser-cams" introduced nearly a decade earlier as an accessory for the corporation's eponymous electroshock weapon. Its top-selling wearable "Axon" camera (which since 2017 has been the company's name) streamlined the device. Gone were more cumbersome elements such as a separate controller, internal storage system, and LCD screen; video and accompanying metadata would be automatically updated to the cloud and accessible to officers via smartphone apps.[18] The technology has its roots in the company's controversial Taser electro-shock weapon, for which it was formerly named. In 2006, the company marked the "Taser-cam" to police departments desiring more mobility in recording their encounters with civilians beyond the restricted perspective of the dashboard cameras or incidental security cameras. The weapon-mounted camera was a canny public-relations move for the company, promising greater transparency with regard to when and how Tasers would be used in the face of growing public concerns about the dangers of these putatively nonlethal "pain compliance techniques."[19]

Brown's brutal killing by White police officer Darren Wilson was followed by weeks of civil unrest and massive public outcry as activists from Missouri and across the nation converged on Ferguson to protest anti-Black police violence. In the wake of these protests, interest in bodycam technology exploded. Protests in Ferguson reignited outrage surrounding the shooting death of another unarmed African American teen—seventeen-year-old Trayvon Martin, two years earlier. While visiting the Sanford, Florida, gated community where his father lived, Martin was fatally shot by a self-styled neighborhood watch patroller. As with the police shooting of Michael Brown, Martin's murder was precipitated by his visual "capture" within the diffuse panopticism of police and neighborhood vigilante profiling. (As seen in the horrific shooting death earlier this year of African American jogger Ahmaud Arbery, these categories are not always distinct.)

Bodycams were prominent in the public narratives surrounding Brown's killing in debates over what exactly forensic evidence and conflicting eyewitness accounts revealed about whether Wilson or Brown himself had been the aggressor. Wilson insisted he had shot Brown in self-defense while the latter was assaulting and threatening him; to the contrary, several witnesses stated that Brown had his hands raised in a gesture of surrender when he was shot and killed, a claim partially corroborated by postmortem findings. These contradictory accounts and the official investigations that soon followed highlighted the fact that Wilson was not wearing a camera at the time, nor had any civilian filmed Wilson's violent assault on Brown. Prior to Brown's death, the Ferguson Police Department did not have a bodycam program. Nor were cameras mounted on the dashboards of their officers' cars. In the public discourses collated on social media by the Ferguson hashtag, the absence of cameras seemed inextricable from the fact that there was even a question over whether the shooting was unjust or justified. Brown's own parents, Lesley McSpadden and Michael Brown Sr., released a public statement that reflected this viewpoint. They asserted that "had Officer Darren Wilson been wearing a body camera, which are being worn by more and more police departments around the country, there would be no questions." Following the grand jury decision that ultimately exonerated Wilson, the absence of body cameras was frequently invoked in public commentary as the reason why Wilson was not brought to justice.

This conclusion deploys popular beliefs regarding the indexical truth value of the documentary image on behalf of the moral and political imperatives of the Black Lives Matter movement. It quickly gained traction in national discussions about police. Just days after Brown's killing, an activist

from Georgia created an online citizens' petition to the White House calling for a federal "Mike Brown Law," requiring "all state, county, and local
police . . . [to] wear a camera." The petition, which soon surpassed the
100,000-signature threshold requiring an official response from the White
House, recapitulated Brown's parents' faith in the body camera's potential
as an instrument of racial justice by claiming that the images it records
"remove all question from normally questionable police encounters."[20]
Regarding the practical implementation of such programs, the petition
said nothing about officer training or department protocols for using the
technology, or how digital storage of and access to these videos should be
regulated.

Popular support for bodycam programs as an antidote to perceived
pro-police and racial biases in the legal treatment of officers accused of
shooting unarmed Black citizens and a deterrent to police misconduct and
abuse soon spread to national policy conversations. Tacitly aligning himself
with the concerns of Ferguson protestors and allies, President Barack
Obama announced, following Wilson's acquittal, that his administration
was asking Congress for $263 million in funding to support the rollout
of new body camera programs nationwide. (Brown's parents themselves
subsequently gathered over a quarter of a million signatures petitioning
the US Senate and House of Representatives to fund Obama's initiative.)[21]
New York City's police department, the nation's largest, pledged to outfit
20,000 officers with cameras by 2019. By the time the 2016 presidential campaign season got underway, criminal justice reform proposals
specifically involving body camera technology were incorporated into the
platforms and public positions of the leading candidates of both major
parties. Just days after announcing her candidacy in April 2015, presumed
Democratic frontrunner Hillary Rodham Clinton, in a keynote speech at
a Columbia University public policy forum, mentioned the tragedy and
unrest in Ferguson by name in her call for federal monies; Clinton added
that body cameras were needed to "protect good people on both sides of
the lens."[22] On the progressive edge of the political spectrum, Clinton's
major opponent for the Democratic nomination, Senator Bernie Sanders,
introduced a platform that revived activists' dream of a "Mike Brown Law."
By demanding policing reforms that specifically recognize "that black
lives matter," his criminal justice reform proposals decried "intolerable
acts of violence being perpetrated by police and racist acts of terrorism by
white supremacists" and called for the federal government to "fund and
require body cameras for law enforcement officers to make it easier to hold

them accountable."²³ GOP candidate Donald Trump promised that federal funding would be made available to all police departments who wished to voluntarily adopt body camera programs, citing the technology's important role in exonerations of officers accused of misconduct.²⁴ Whether to protect officers from false allegations, make encounters between police and citizens safer for all parties involved, or end the police's "war on black people,"²⁵ the body camera had become, in the wake of Ferguson, the candidates' police reform panacea of choice.

As Howard M. Wasserman argues, the popular and political embrace of body cameras as a solution to the complex issues involved in the problem of police misconduct generally and against people of color specifically illustrates a tendency in contemporary American policymaking to pass laws in response to "moral panics." Wasserman uses the term to describe the sudden intensification of public fears about social phenomena believed to threaten societal values or order. Examples include "laws in areas such as child sexual abuse, child pornography, fetal protection, financial regulation, and illegal drug use [that] have been criticized as overreactions to moral panics, often because the laws represent solutions that are unsuited or wildly disproportionate to the actual problem, although sold to the public as an easy cure-all."²⁶ In the discursive community of #Ferguson, unlike the actual community of Ferguson, Missouri, the threat to the social order was not structural racism, racial disparities in the criminal justice system, or racist policing, all of which the tragically emblematic death of Michael Brown threw into stark relief. Instead, within the majority White cultural mainstream that would soon come to embrace the call for technologized transparency in policing initially sounded by police accountability activists, the moral panic over "Ferguson" was engendered by the protests themselves and their perceived threat to the dominant social order.

For the Ferguson protests foregrounded a newly mobilized Black political culture oriented by emotions of "grief, anger, fury, rage, terror, and exasperation as incidents of police violence against unarmed or legally armed black people continue month after month, and even those captured on video and circulated online fail to result in indictments of the officers involved."²⁷ Debra Thompson notes that the Black Lives Matter political movement that coalesced through demonstrations in Ferguson and around the nation has specifically repudiated anti-Black "respectability politics." Thompson describes the racialized ideal of respectability as "a constitutive element of white supremacy" that makes African American conformity to putatively "white" cultural norms concerning behavior, appearance, and speech a qualification of civil personhood and political recognition. This racist politics of respectability routinely features in how African American victims of police violence are held up to public scrutiny in the wake of their deaths. A marijuana habit, a penchant for wearing black hoodies, a previous conviction, recent meth-amphetamine use—biographical details like these are used to construct a posthumous counternarrative of the victim as criminal.

For Thompson, the moral panic catalyzed by the Ferguson civil unrest and #Ferguson activism nationwide reflects how White society tends to frame Black rage as not just inappropriate but undemocratic "because it explicitly challenges the pervasiveness, durability, and applicability of the American Dream, itself premised on the liberal myth of inevitable progress toward a more egalitarian society."[28] In explicitly using Black rage "to challenge white supremacy," Black Lives Matter's politicization of African Americans' and their allies' grief and anger in order to instigate public protest and acts of civil resistance is, Thompson concludes, received as particularly problematic.[29] What makes the federal police body camera legislation now under consideration and initially proposed by Obama characteristic of the moral panic paradigm evoked by both Wasserman and Thompson? As Wasserman argues, body camera laws' resemblance to other juridical dis-courses stemming from moral panics lies with the questionable legal value of videos as evidence and a general disregard for issues of implementation.

Funding for cameras alone does not ensure that they are used appropri-ately or even used at all. In addition to the costs of equipment purchase, departments would have to contend with even more expensive matters of training and long-term data storage.[30] Indeed, the city of Ferguson is a case in point: at the time of Brown's shooting death and in the civil unrest that followed, the police department already owned two body cameras but lacked the money to pay for operationalizing the technology and for video storage. Too, there is the additional matter of public access to the videos, which seems a critical dimension of the post-Ferguson call for transparency in policing. At this writing, half of all states have passed or pending laws restricting public access to body camera video and all but New Hampshire have existing laws exempting police from public records requests that may affect access to these videos, too.[31] Even where public access is not de jure blocked, the videos may be de facto inaccessible, given that the cost of preparing videos in response to a public records request can run as high as $18,000.[32]

***

The threads of my discussion converge in the 2015 police shooting of Samuel DuBose, the legal exoneration of his killer, and the central message of the protests spanning these two moments. "YOU LIED! BODY CAMERAS DON'T." In urgent, bright red letters in all caps, emblazoned on a large piece of black poster board, this is the message that Audrey DuBose addressed to her son's killer, on a large hand-held sign she brought with her to protests and public rallies, media interviews, and marches in in the months leading up to his November 2016 trial on charges of murder and voluntary manslaughter.[33] On July 19 of the previous year, University of Cincinnati police officer Ray Tensing shot Samuel DuBose in the head at close range during a traffic stop while patrolling an off-campus neighborhood. At the time of his death, DuBose was sitting in the driver's seat of his car. According to Tensing, he shot DuBose in self-defense after the motorist had effectively weaponized his vehicle by attempting to drive off with the officer's arm caught in his car.

Pitting the officer's claims against video footage that bears horrifyingly intimate witness to the killing, Audrey DuBose's protest poster gestures, in its semiotic linkages, toward the broader political context in which Tensing was indicted for murder. Beneath the accusation "you lied" running across the top of the poster is an enlarged copy of a photograph of the killer in his police uniform. The image of Ray Tensing is an official department photograph, and DuBose's choice to use this picture, rather than an image of Tensing taken from the news or social media, emphasizes his status as an agent of the state who committed this act of violence in the course of his official duty. Below Tensing's face, Audrey DuBose included another photograph, one that endows the racial semiotics of the event—a White officer shooting and killing an unarmed Black citizen—with an additional semantic inflection beyond the politically symbolic polarity of predominately White (and White-protecting) officers and African American communities targeted for state violence. This second image is not a photograph of "Sam," whose smiling portrait could be seen on T-shirts worn by members of the DuBose family and their supporters in public demonstrations in the days leading up to Tensing's trial on charges of murder and voluntary manslaughter. On the protest poster, the "face" that confronts the officer's is the lens of the body-worn video camera that recorded the shooting from Tensing's perspective as it took place. The visual opposition between Tensing's racially White face and the camera's black plastic casing frames the bodycam as a proxy for the Black witnessing public subjecting White police misconduct to counter-surveillance.

To illustrate the dictum that "body cameras don't [lie]," Audrey DuBose included a simple stock photograph (presumably an image originally produced for marketing purposes) that features two common models of police body cameras, like the one used to record the shooting. The visual composition of Tensing's photograph is structured by a stark chromatic dichotomy between the phenotypical lightness of Tensing's face and the blackness of his uniform and of the photograph's background. The foreground/background arrangements of the bodycam photograph present an inversion of this pattern, as its black objects appear to "float" in empty white space. The all-white background emphasizes the opaque blackness of the cameras' blocky plastic casing and the dark luminescence of their unblinking, all-seeing "eyes." The juxtaposition of these two photographs privileges the technologized vision of the body camera as morally authoritative. "Tensing may be a liar," we are meant to conclude, "but his camera is not."

DuBose's protest poster and its collage aesthetics that confronts the police officer's alleged lies with his body camera's seemingly inexorable truth provide a synopsis of a dominant cultural logic that has conjoined the demands of racial justice to the increased techno-visibility of American policing. For the DuBose family, the prosecutors, and many within the witnessing public constituted by the video's public release, the bodycam footage presented incontrovertible evidence of murder. Yet it was ultimately unpersuasive for juries tasked with adjudicating Tensing's guilt. The 2016 trial ended with a hung jury. A second trial, in June 2017, had the same result, and on July 24, 2017, Hamilton County (Ohio) Judge Leslie Ghiz dismissed all charges against Tensing.

In spite of the state of Ohio's failure to hold Tensing criminally responsible for DuBoses's death, the outcome of the case left unshaken a broader public belief in documentary visuality's capacity to disrupt the racialized structures of perception expressed in norms and practices of "racial profiling" through which the state constructs and maintains the precariousness of African American lives and normalizes the traumas of Black social and biological death—so deeply ingrained are cultural presuppositions about the probative legal value of the body camera's instrumentalist aesthetics. It is of no matter that the discourse of documentary visuality that informs the idealization of the bodycam is itself shaped by racialized structures of perception. Caren Myers Morrison summarizes the much more ambiguous legal reality that assumptions about the "transparency" of the encounters depicted on video recordings of police violence hide:

> When these recordings are used as evidence in police use-of-force cases, the task of the fact finders is to divine whether the police officer's actions were

"reasonable" under the Fourth Amendment. Due to a confluence of factors, including a highly deferential legal standard and the biases that affect any viewer, video recorded by police may benefit the police more often than not. Grand juries might fail to indict, or petit juries vote to acquit, and the video images, which seemed to promise clarity, become another source of confusion.[34]

When it comes to bodycam videos, more than with dashcam or bystander videos of violent police encounters, Morrison further observes that "the legal standard melds with the perspective of video."[35]

By the time of Tensing's November 2016 trial on murder charges, nearly a third of the nation's 18,000 police agencies had begun using the cameras, either as standard procedure or in departmental pilot programs.[36] The trial of Samuel DuBose's killer presented an opportunity for body camera technology to make good on its ostensible promise as an agent of racial justice and social change.[37] University of Cincinnati's campus police force was an early adopter of bodycam technology amid the post-Ferguson moral panic earlier described by Wasserman and Thompson. The trial took place following a spate of non-indictments, acquittals, and convictions on lesser charges in similar cases. Unlike in this trial, visual evidence in these prior, failed efforts to seek criminal convictions in what mainstream discourse euphemistically terms "officer-involved fatalities" was either nonexistent or limited to the more distant, attenuated perspectives of police dashboard-mounted or security cameras or video from bystander cellphones, all of which defense attorneys are quick to discredit.

That Samuel DuBose's death was captured on camera—in graphic visual and audio detail—from the first-person perspective of the police officer who shot him suggested that the legal scales were dramatically tipped in favor of the prosecution. At least this was the viewpoint broadly shared by the victim's family members, the prosecutors in the case, and by many members of the public. That there was a murder charge at all, in addition to indictment for manslaughter, was held up by the DuBose family and their supporters as proof of the body camera's transformative implications of ending police violence in African American communities. At a pre-trial rally, Audrey DuBose confidently told a crowd of supporters that "this situation here, with my son, is going to make a change."

Apart from what was at the time a novel presence of video documentation, the occasion of DuBose's death conformed, in its particulars, to what Steve Martinot describes as "the structure of the form of an event that has reached epidemic proportions in the United States": an officer pursues or detains an unarmed Black male, shoots him dead, and then claims to have used his gun only in self-defense, believing his own life to be imperiled.[38] This structure could be further elaborated to include the victim's posthumous criminalization and the officer's exoneration, often without even being brought to trial or even before a grand jury. DuBose's mother and brother might well have these events in mind when, in a *New York Times* documentary, they affirm for the journalist in an interview their shared conviction that, without footage from Tensing's body camera, not only would there be no case against Tensing, there would be no "story" compelling enough to capture national media attention. Perhaps his mother is thinking specifically of the grand jury's decision the previous year not to indict Darren Wilson, the former Ferguson police officer who killed Brown, when she tells the *Times* reporter, "There's so many times, where we saw this happen where there's no justice; if it wasn't for the body cameras, Ray Tensing wouldn't have been charged at all." Likewise, DuBose's brother Aubrey evokes both Wilson's non-indictment and the acquittal of George Zimmerman, Trayvon Martin's killer, in comments that credit the bodycam video of his brother's shooting with transforming public perception of the killing from a virtual non-event into a singular news story: "Without the body camera, there wouldn't be nothing. You [the news media] wouldn't be here; *there wouldn't be no story*. He'd be a hero, my brother would be a criminal—cut and dry."

Aubrey's comments point to the powerful semantic influence of "the structure of the form of [this] event" in producing an *a priori* account of his brother's death as a justifiable use of what is euphemistically known as police "force." What keeps *this* event from being slotted into a hegemonic narrative about Black criminality is the very different story the video itself presents. Protests following DuBose's killing and rallies of support in the weeks prior to Tensing's trial resounded with claims that it is the camera that carries the "story" of how DuBose was killed. Also shown in the *Times* documentary is a woman speaking through a bullhorn to a large crowd in downtown Cincinnati that gathered to protest DuBose's killing: "If there was no camera at the scene, the media would have taken the word of Tensing and the other officers." Applause breaks out even before she has finished the sentence.

Tensing's body camera footage was surely crucial in securing an indictment against the officer. However, even though the prosecution in this criminal case pointed to the video as unmistakable evidence of Tensing's guilt, Tensing and his defense team were just as adamant that the video proved his innocence. Tensing pled not guilty to the charges of murder and voluntary manslaughter, claiming that the use of "deadly force" was justified on the basis of the threat his actions posed to the officer's own life. The bodycam video, released to the public ten days after the shooting, in response to widespread demands both in the media and on the streets, was widely framed in its public media reception as contradicting Tensing's account of the events. Moreover, during the trial itself, expert analysis—including that of an impartial third party contracted by the state—supported the prosecutor's case by disputing two crucial points in the self-defense narrative. Tensing testified that his sleeve was "caught" in DuBose's car door and that the vehicle was moving at the time of the shooting, which led him to fear he might be grievously injured or killed by being dragged by DuBose's speeding car. Expert analysis of the bodycam video found that neither claim was true: the sleeve was *not* caught and the car was *not* moving.

Tensing's lawyers, however, continued to insist in public statements both following the video's release and throughout the trial and retrial that the video supported their client's story. Significantly, this story took shape in relation to the video itself: Tensing spoke to the detectives investigating DuBose's death only *after* he himself had had a chance to review the video. (The accused officer's access to this video represents an explicit violation of his department's stated policy.) What he "saw" when he viewed the video was precisely what he "saw" within the context of the lived events leading to his killing of Dubose. His defense rests on evidence that is not nakedly visible in the bodycam footage but seems quite literally to reside in the eye of Tensing as the video's authoritative beholder.

In its capacity to support in court two irreconcilable interpretations of what happened in the seconds leading up to DuBose's death, the Tensing bodycam video points to the technology's hermeneutic malleability. This stands in stark contrast to the DuBose family's expectations, which were keyed to common understandings of body cameras as instruments of transparency and police sousveillance. Such conceptions of the technology miss how the discursive and material construction of "perspective" in the legal framing of bodycam and other police videos is even more significant to the outcome of such cases than the indexical "thereness" of video imagery.

The acquittal of Ray Tensing registers how the form of spectatorship produced by bodycam videos readily conforms with the identificatory structures and haptic ontology of first-person shooter videogames. Citing the names of the two most widely known first-person shooter games, Dexter Thomas, writing in the *Los Angeles Times*, analyzes the nearly thirty-minute video capturing DuBose's shooting death, its prelude and immediate aftermath,

as formally mimicking videogames as a narrative genre: "Just as in 'Call of
Duty' or 'Halo,' we never see the officer-protagonist's face or full body—only
his hands, as he gestures or holds his weapon. We see the story progress
from a simple traffic stop, and get just enough time to feel at ease with
the calm conversation. So when the single shot is fired, the viewer may be
jolted out of the narrative to wonder, why did *I* do that?"[39] Unlike dashboard
mounted cameras or citizen-wielded cellphones, which present third-person
perspectives that cast the viewer as witness or bystander to violent police
misconduct, the bodycam interpellates the viewer as a direct participant
in the scene—as the killer himself. Published just days after the video's
release, Thomas's interpretation is optimistic about the ethical effects of
being "jolted" by the video's structure of viewership. The question "Why
did *I* do that?" comes, in his analysis, from a place of revulsion and horror
at being made to be the killer and forces a visceral, moral engagement with
the scene of Black death as a situation of (White) viewer complicity. The
analysis implicitly favors the moral occasion of the bodycam's immediacy
over the distanciated "looks" produced by dashcam and civilian bystander
videos. In Thomas's reading, the latter are more likely to make the mainstream (and, again, implicitly White) viewing audience inured to spectacles
of African American violent death.

For Thomas, the mediated "I" in this formulation is disruptive and
potentially effective, from a social justice perspective, because it is specifically interior to the spectator rather than lodged within an identification
with either of the social actors portrayed within the video. While DuBose's
objectification as the embodied subject at the center of the video's field of
vision prevents him from being, in structural terms, the spectator's object
of identification, neither is the officer, Thomas argues, because the point
of view that organizes the video invites the viewer to see in his hands the
image of her own.

Morrison also describes the experience of watching Ray Tensing's bodycam's video as akin to playing first-person "shooter" videogames but arrives
at a different conclusion, regarding the identifications at play. "With just
a small leap of imagination," she writes, "I can believe that I'm seeing my
own hands, my own gun."[40] In first-person, point-of-view video games, the
structure of identification draws the videogame player into the role of the
shooter, a dimension of the gaming experience that undermines Thomas's
argument about the potential for an ethical and just witnessing via the
bodycam video. Thomas's account implies that the officer is somehow not
onscreen; however, in the video we do see Tensing as an embodied subject,

even if it is just his arms and his gun that appear. His hands' appearance on the screen, where the viewer's own are not, invites us to see him, in narrative terms, as the video's protagonist. In this way, Morrison argues, the body camera video, unlike video evidence framed from a third-person perspectives, draws the viewer into "the closest identification possible with . . . the police officer."[41] "For a judge or juror trying to determine whether the 'protagonist' acted reasonably, the perspective of police video puts at least a thumb on the scale of sympathy for the officer. We are threatened by the suspect, we are chasing the running man, we are jostled and surprised by sudden violence. The factfinder, then, is not only asked to evaluate whether the action was reasonable, but also to evaluate it from a police perspective that the video invites her to share."[42] Considered in this light, the arming of American police departments with body cameras might make it even less likely that a death like DuBose's would end in convictions, compared to the third-person videos and their more open, ambivalent structures of identification.

Even while visual technology makes policing more visible, the very intelligibility of police violence *as* violence—as opposed to an instrumental use of force in keeping with police work's professional protocols—is constituted, enacted, or suppressed by the often racially discriminatory adjustments of legal norms concerning "reasonable force." These norms converge, in the juridical context, with the importance of officer testimony, the institutional relations between prosecutors and the police, and the protocols for grand jury hearings. Especially in relation to jurors' and judges' understanding of "reasonable force," as Thompson notes, research on racial attitudes fulsomely demonstrates that "African Americans are perceived as more aggressive, more dangerous, physically stronger, and less prone to feeling than white people."[43] Part and parcel of ideological anti-blackness in the United States, such presuppositions have significant implications for how bodycam footage functions as a genre of legal evidence. I would argue that it is what many viewers describe as the "intensity" of bodycam footage that is especially problematic in the courtroom, given the capacity of these videos to, as Morrison notes, "heighten the sense of danger to the officer."[44]

***

I have described how the history of photography—especially, its use by the state as a visual apparatus for social control—informs the troubling representational regimes of bodycam technology. I now turn to analyses of visuality and surveillance in the history of New World slavery as a further elaboration of the racial discursivity of the bodycam. Under slavery, the systematic visual reproduction of embodied blackness as the symbolic, biopolitical substrate of racial capitalism required controlled spectacles of suffering alongside the *visual dissimulation* of slave suffering. According to Saidiya Hartman, as it was institutionalized in the United States, chattel slavery was constituted in and defined by "the spectacular nature of black suffering and, conversely, the dissimulation of suffering through spectacle."[45] This formulation resonates with the material-discursive practices involved in the video documentation of anti-Black police violence. Racial spectacle—including forms of visual consumption and commercial visuality—was integral to chattel slavery's organizing social, economic, and legal practices.[46] Hartman's analysis, which draws from extensive archival research and enslaved individuals' own narratives, demonstrates that spectacle was essential to reproduction of capital in the form of human flesh. Slave marketplaces drew their customers and secured top prices for their human commodities by creating a carnivalesque environment—including food, drink, leisure entertainments—that showcased coerced displays of leisure and pleasure (such as singing performances), of strength and fitness, and of pleasing personalities and good humor—all public ritualized expressions of White mastery. Enslaved individuals were expected to participate in their own marketing, under threat of physical violence, helping to sell themselves by telling prospective buyers the labor they could do. In this Rabelaisian nightmare, writes Hartman, "the stimulating effects of intoxicants, the simulation of good times, and the to-and-fro of half-naked bodies on display all acted to incite the flow of capital."[47]

The discursive and state practices involved in the production of the enslaved body as a spectacle reveals visuality as a technique of violence and subjection. This dynamic that might be obscured by the contrast between a more obviously spectacular practice of violence—the "thirty lashes" that one account notes was promised to the "sullen" slave who fails to sell herself enthusiastically—and the quotidian violence of "half-naked bodies on display." The slave market distributes the categorical distinctions between Black and non-Black and states of unfreedom and liberty across space. Within the larger institutions of slavery, these spatialized taxonomies are also distributed in time, through ritualized repetition. In this sense, the repetitions of circulation within a digital public sphere animated by modalities of refreshing, hash-tagging, bookmarking, linking, reposting, and sharing, work to "solidify," as Christina Sharpe notes, "and make continuous the colonial project of violence."[48]

For Nicholas Mirzeoff, the modern-day regime of racialized surveillance in which the video document of police encounters with citizens colludes with practices of racial profiling in policing and the law looks to the project of visuality founded in Atlantic world plantation slavery. This is the "plantation complex of visuality" whose vestiges organize the field of representation that regulates how Black suffering appears and how it is seen.[49] Mirzeoff describes the "plantation complex" as "a system of visualized surveillance"—including mapping, natural sciences, and slave discipline—that mediated sovereign authority through the visualized techniques of power the enslaver held over the enslaved. Collectively comprising the plantation manager's "oversight," these techniques spatialized concepts *liberty* and *unfreedom* and oriented the overseer's power over the body of the slave around a specific delimited gaze that was at once scientific, spatializing, and legal.

It is thus that, in the context of the plantation, the force of race law and the racialization of law were institutionalized as the power to look and control looking. As with Hartman's analysis of the ruse of slave pleasure, a spectacle that mediated both slave suffering and its dissimulation, Mirzeoff describes the panoptic plantation complex as virtualizing "oversight" as a structure of violent repression. The implication is that, for the institution of slavery, visuality operated as the power to control reality: "Under slavery, the enslaved were forbidden to 'eyeball' the white population as a whole, an injunction that was sustained throughout the period of segregation and is active in today's prison system."[50] This prohibition helped create a scopic regime for the administration of authority in slavery that subsequently shaped the asymmetrical terms of corporealization within racial modernity.

Film scholar Richard Dyer captures what I mean by "racial modernity" and its disparate productions of dis/embodiment in his analysis of the different positions that "whiteness" and "blackness" inhabit in the "imagery" of race as ideology. Blackness, as the locus and baseline of race as a visual discourse, is reduced to the corporeal, while whiteness in modern visual culture is essentially disembodied, a formulation that leaves "the nonwhite body prey to the promptings and fallibilities of the body."[51] In contrast to what I earlier invoked as the fragmentary glimpses of embodied (police officer) subjectivity screened by the singular technology of the body camera, in the broader representational field, "embodiment" is all fallibilities and only prey. What Dyer means by "the body" is specifically one that is capable of subjugation within and through public exposure. This is the body that exists in public, as Lauren Berlant writes, as "an obstacle and not a vehicle to pleasure or power."[52] In other words, within the plantation matrix described by Mirzeoff, interdictions that designated the White body as that which cannot be looked at made whiteness, as that which cannot be seen, discursively invisible. Social media uploads and streaming of police brutality videos enact a desire to undo that ideological formation by "eyeballing" White officers, thereby making visible systemic racism and the structures of power that reproduce racial "whiteness." Unlike the plantation panopticism identified by Mirzeoff, the diffuse, distributed nature of power within the present day's "societies of control" does not present clear lines of sight by which counterhegemonic eyeballing—which is to say, sousveillance—might disrupt, redirect, and dismantle the operations of racial hegemony.

In the post-panoptic present, videos of police brutality ultimately accumulate as a collective document of how Black embodiment is produced juridically-discursively in and as the nightmare obverse of modernity's signature spaces of individual freedom: the city street, the automobile, the individual body. Online, the terrifying bodycam footage depicting the shooting of Samuel DuBose at point-blank range coalesces with surveillance camera, dashcam, and cellphone imagery of the dying moments of Philando Castile, Eric Garner, Terence Crutcher, Oscar Grant, and Tamir Rice. Depicting the deaths of motorists, a mass transit rider, a pedestrian on a street corner, a child at a public park—this archive invokes antebellum distinctions between spaces of freedom and captivity in and as the spatialized declensions of Black unfreedom in the urban street, the vehicle, and finally the Black body itself.

If the legal treatment of police shooting videos reveals the bodycam's promises of accountability to be a ruse, one of the most important effects of that ruse is that it dissimulates the alarming *non-transparency* of the broader incidence of civilian deaths at the hands of police. While the creation of such an agency is one of the provisions of the proposed George Floyd Justice in Policing Act of 2020, there is no single federal office that comprehensively monitors the number of civilians killed by police each year. There exists no official statistical outline of what is increasingly and rightly described, in biopolitical terms, as an epidemic. As D. Brian Burghart, one of the journalist activists who has started tracking this information himself, observes, "The federal government tracks anything that matters . . . the number of shoes sold, rainfall in Death Valley. The fact that they weren't collecting this information suggests that it just didn't matter."[53] There are important implications here for the racialized visuality that mediates the body-camera video in the courtroom. Police killings of unarmed African Americans cohere as a cultural trope but not as a social fact.

Data collection serves as a privileged form of official state knowledge. It also constitutes a scopic regime by which structures of power and systemic inequality are made visible. Statistical information about police shootings of civilians, including data about victims' race and ethnicity, enable single events to be perceived as comprising part of a pattern. The state's refusal, in effect, to "look" at this phenomenon through data collection is a constitutive feature of the post-panopticism that inscribes the systems of surveillance in which these deaths occur. Statistical invisibility restricts the structures of perception through which the law "sees" anti-Black police violence. Instead the critical task of making anti-Black police violence legible falls to visual technologies: the body camera, dashcam, and video-recording smartphone. These technologies individuate the events they depict. Their images frame, for audiences of juries, judges, and witnessing publics, the actions they depict in terms of either an individual bad actor or an exonerated officer acting in compliance with police use-of-force protocols. Both paradigms obscure the role of systemic, institutional racism. In the absence of any official data by which to be able to "see" and scrutinize police violence as a disciplinary dimension of what David Theo Goldberg terms "the racial state," documentation of Black death and suffering—which the body camera threatens to make only even more routine and more graphic in its grisly atomized details—will simply accumulate on the digital shoals of the contemporary racial interface as mere assemblages, without any claim on the project of visuality that organizes the spectatorial relations that condition and control legal recognition. Suggesting parallels with the plantation complex model identified by Mirzeoff, the federal government's refusal to admit the subject matter of police killings of civilians within the "gaze" constructed by its technologies of bureaucratic informatics evince a sovereign power over the "reality" that police videos are tasked with mediating.

The lacuna of data constrains the bodycam's potential for prosecuting police misconduct, including murder and manslaughter charges. Its evidentiary weight is undercut by the lack of official state knowledge about when, where, why, how, and with what frequency police take the lives of civilians. The statistical invisibility of police shootings in general and African American fatalities occludes a broader legal understanding of the institutional cultural context for deaths such as Samuel DuBose's. In a context in which the state refuses to admit to public view a "full picture" of police killing of Black civilians, the video in its juridical construction as witness to racial violence can only see far too much or not nearly enough.

In order to see the video's indexical referent, academics, policy analysts, activists, and the media rely on sources of data currently produced by journalists and activists.[54] Two news organizations, the *Guardian* and the *Washington Post*, have compiled databases that consult search engine feeds for their data on police shootings. Websites such as Burghart's *Fatal Encounters* and the database *Killed by Police* are not affiliated with news organizations. They rely on media reports and crowd-sourcing to develop a national picture of deadly police shootings. The *Post's* database, which began in 2015 as a direct response to the events in Ferguson, focuses on fatalities involving guns in the years since the project began.[55] According to the *Post's* accounting, fatal shootings by police have remained at a fairly constant number since the organization's project began in 2015. That year saw 991 shooting deaths recorded. There were 963 in 2016. The cumulative total of fatal shootings through November 22, 2020, is 5,783.

These organizations' figures reveal that, post-Ferguson and the bodycamification of American policing, African Americans remain disproportionately targeted for death. Black males account for only six percent of the total population yet constitute nearly a quarter of those shot and killed by police since the beginning of 2015.[56] They form an even greater portion of the total number of shootings of unarmed civilians—roughly one-third. Considering the legal and cultural implications of police body cameras, many critics compare the images these cameras make, in the context of fatal police-civilian encounters, to the historical practice of lynching photography.[57]

This history also encompasses visual activism, on the part of anti-lynching crusaders, that provides a precursor to the contemporary project of countervisuality constituted in the social media distribution of images of police shootings. Anti-lynching crusaders with the NAACP used these images as evidence of oppression and atrocity, in an effort to galvanize political recognition and action to put an end to the tacit legal sanction given to anti-Black lynching. Leigh Raiford terms this archive of counterhegemonic deployments "anti/lynching photography"—the virgule denoting how the representation of lynching is not fully detachable from the violent spectacle in which it originates. As Raiford argues, this archive "has been central to the recounting and reconstitution of black political cultures throughout the 20th century. From the usage of lynching photography in pamphlets by turn of the century antilynching activists, to posters created by mid-century civil rights organizations, to their deployment in contemporary art and popular culture, we can see how this archive has been a constitutive element of black visuality more broadly."[58] (Within this broader history of Black countervisuality, Raiford cites as particularly politically effective the creation of an outraged witnessing public around the postmortem images of the brutally murdered fourteen-year-old African American Emmett Till, images published in periodicals and newspapers at the insistence of his mother.)

Like the police-worn camera, the lynching photograph constructs spectatorship in terms that can be analogized to the gaze of first-person shooter videogame. In lynching images, perspective is framed via a weapon. However, the weapon in this instance is not seen within the frame for it *is* the frame. Photography participated in the lynching spectacle, not to document but to administer and intensify the violence against the Black body and to amplify the spectacle's reach through the late nineteenth and early twentieth centuries' versions of "social" media: the postcard and commercial photograph. Many of the surviving photographs reveal the photography studio's trademark, suggesting that lynching imagery was used to publicize their trade. It was common for them to be produced as postcards, enabling White witness-participants to keep a souvenir of the event or those unable to be present to participate virtually. Lynching photography documents the camera's use as a weapon of violence against the Black body. It also marks continuities in the aftermath of slavery with the scopic regimes described by Hartman regarding the spectacular commodification of blackness as a tool of surveillance and social control. An amplification of the spectacle, extending its symbolic reach, the lynching photograph produced and sold as a souvenir image enabled the White supremacist terrorism of lynching violence to be enacted through and across visual culture, extending the terrorizing reach of lynching violence beyond the localized space and time of the event.

It was the camera—as much as the specifically ritualized forms of violence used by lynch mobs—that rendered lynching a "social ritual," one that involved old and young, men and women alike. In elaborating the "spectacular" qualities of lynching violence, historian Leon Litwack notes how when these extrajudicial executions were planned in advance, newspapers announced "the time and place of a lynching, special 'excursion' trains transported spectators to the scene, employers . . . released their workers to attend, parents sent notes to school asking teachers to excuse their children for the event, and entire families attended," often packing picnic meals.[59] The use of the camera to commemorate these communal orgies of violence "testified to their openness and to the self-righteousness that animated the participants."[60]

Beyond its representational function, lynching photography participated in the violence it depicted through the practices of reproduction, distribution, and consumption by which these images circulated *as commodities* in the popular public sphere to produce and maintain a civic culture that tacitly sanctioned the extralegal punishment of African Americans for any perceived threat to the racialized social order. The fact that images of lynching primarily entered American social life via the popular public sphere as picture postcards, a mass-produced and distinctly intersubjective commodity, suggests another set of historical coordinates for thinking about the fate of visual evidence of what activists organizing around anti-Black police violence deem modern-day lynching violence. In many ways, the circulations and transmissions of the picture postcard—a communication technology that dates to roughly the same period as the advent of anti-Black spectacle lynchings—might be read as rehearsing the circulations and transmissions of contemporary digital social media as its visual form and as networked structures intermix private intimacies with public displays. Unlike letters conveyed in envelopes, postcards compromise the privacy of their addressee. They also belonged to a commercial public sphere more broadly, beyond the semiprivate networks of the postal route, in that these commodified representations of Black death and suffering were "sold in local establishments and sent to family and friends." The scope of communal participation in lynching spectacle expanded, as Amy Louise Wood notes, "to include sympathizers in other towns and regions. The broad circulation of these images further transformed even 'private' lynchings into rather public spectacles, produced for collective consumption."⁶¹

The historical archive of lynching photography shadows a digital archive that is now proliferating in the era of the police body camera and smartphone sousveillance. We should consider the social activist and legal projects attached to this new technology in relation to the historical production of the mutually codetermining and constituting relations between the law, race, and the visual that these newer media technologies at once enact and mystify. As with scholarly attention to the postal and commercial circulation contexts that constitute lynching photography's media frames, the networked digital circulation, reduplication, and archiving of videos of African Americans killed by police officers transform these images into spectacles of suffering whose very spectacularity can be seen to negate the reality of the suffering and hence their affective and ethical charge for viewers. Observes Black journalist Nehal El-Hadi, the deaths depicted in these videos "acquire a new permanence" via the digital reduplications of *refreshing* and *reloading*, which subsumes the videos' meaning. The moral imperative of witnessing in this instance might lie in the refusal to "click" on the "uploaded videos of the moments where Black men and women lose their lives."

El-Hadi's uneasiness about the visual and social logics of the digital interface—social media, especially—that screens Black death in order to incite political action in the present era of antiracist and police protests extends to the "live" feeds of bystander video, which might be seen as circumventing the ethically problematic practices of image consumption that, she argues, "freeze" these mediated deaths as racial spectacle. The 2016 police shooting death in Minneapolis of Black motorist Philando Castile was livestreamed by his girlfriend, who sat in the front passenger seat of his car, via her Facebook page. A reluctant, unwitting virtual witness to Castile's final moments, El-Hadi encountered the video via Twitter's "autoplay" function that automatically plays video in users' feeds when they scroll over them. In contrast to videos of deaths uploaded after the fact, the live video positioned El-Hadi, as a viewer, in "temporal proximity" to his death, positioning her as mute witness, unable to intervene or change the inevitable outcome of what was playing out on her screen. After the event, the more generalized scene of social media spectatorship, in El-Hadi's description, disrupts this affectively charged sense of proximity, ceding the video to the digital medium's capacities for temporal flattening. What soon becomes lost is the video's putative and politically and possibly legally efficacious "thenness"—to riff on Rosalind Deutch's term ("thereness") for photographic and filmic indexicality. This loss is not registered in the surface of the image-text. Instead, the immediacy signified in the video's creation becomes, in the wake of an expired, now archived liveness, a *spectacle* of immediacy.

The ongoing, potentially interminable circulation of these images and videos carries them beyond the temporal limits of juridical processes and the local, community-level moments of crisis through which their meanings are initially framed. These sousveillance and bodycam videos continue to reside on the Internet and circulate via social media long after their testimonial function (in the form of legal evidence and/or on behalf of public social justice work) has been exhausted. Online, they remain available to memorialize and incite yet also, as El-Hadi suggests, to trigger and retraumatize. Even more troubling is how these images, in their digital afterlives, serve as well to routinize and banalize.[62]

As Christina Sharpe observes, "These images work to confirm the status, location, and already held opinions within dominant ideology about those exhibitions of spectacular black bodies whose meanings then remain unchanged."[63] The dissemination and reduplication of these images online robs them of their singularity as documents of the event of an individual's death. As digital assemblages are the principal form that historical memory assumes online, these video images are also subject to more troubling decontextualizing and recontextualizing, recuperations and acts of remembrance, such as we see in the recent proliferation of KKK and even lynching imagery in the visual vocabularies of the mainstream White public sphere. Sharpe calls this to mind in her comment that, "as far as images of Black people are concerned, in their circulation they often don't, in fact, do the imaging work that we expect of them." Even when images of Black suffering are publicly "framed in and as calls to action or calls to feel with and for violences enacted on black people," their circulation "does not lead to a cessation of violence, nor does it, across or within communities, lead *primarily* to sympathy or something like empathy."[64]

Sharpe's critique provides an analytic framework for reading El-Hadi's account of her own resistance to interpellation by traumatic witnessing through the seemingly permanent position of these videos as, in their collective, cumulative presence, the generic afterimage of dominant racial ideology. As its images increasingly enter into a virtual archive of the modern-day equivalent of lynching photography and its atrocity spectacles, the police-worn bodycam is both an analog of smartphone sousveillance, police dashcam video, and public and privately controlled security cameras, all of which have also documented Black death—and a unique media form, due to its distinct formal, institutional, and legal structures of perception. These structures literalize the perspective of dominant ideology that already, via the discursive frameworks of online and news media consumption, inscribes images otherwise believed to witness and demand justice for Black suffering, as Sharpe argues.

The Internet confers a deep horizontality to the sense of time as it is experienced in public culture. Thus, the livestreamed smartphone video of Castile's death acquired through its media transmission an aura of un-liveness as it was immediately, automatically, converted from "live" to "recorded." Too, "un-liveness" registers the recursivity that surrounds this content, in particular, given the spectatorship constructed by this media platform. Viewers who encountered the live feed of the Castile video witnessed the event of his death simultaneously with its filming and public mediation; at the same time, they were unwitting participants in structures of spectatorship previously established around the mass circulation of videos or photographic documentation of police killings of unarmed African American civilians. In the case of these earlier non-livefeed videos, their circulation occurred *after* the deaths they document and continued in the aftermaths of police acquittals. Because of the established structure of spectatorship, even while it was livestreamed, the video of Castile dying of gunshot wounds in the driver's seat of his vehicle conveyed the inexorability of his death, *as if he were already dead*, and the inexorability of justice denied, *as if his killer's acquittal was already assured.* (Indeed, Castile's killer, who shot him seven times at close range as Castile reached for his driver's license in compliance with the officer's orders, was ultimately acquitted of second-degree manslaughter and two counts of dangerous discharge of a firearm.)

In elaborating these dynamics of race and representation, which are constitutive for the visual and social logics I am formulating as racial post-panopticism, Sharpe and other scholars invoke a viral media representation of Black suffering and police brutality that serves to bridge the history of lynching photography and the present-day phenomenon of social media circulations of the present and the home video recording of four White members of the Los Angeles Police Department beating Black motorist Rodney King. In its original circulation, the video was shown repeatedly on the nightly news in 1991 and again, in 1992, during the officers' trial. The video, shot by a citizen bystander, called into being—at least momentarily—a feeling public mobilized around the issue of anti-Black police violence. The image quality is poor—grainy and silent, shot from a distanced perspective. The practices of systemic brutality and racially motivated motorist profiling the video depicted were not "news" to African Americans in Los Angeles, nor elsewhere in the nation. But because King's beating and arrest were witnessed by George Halliday's home video recording camera, they became a "[news] story" in the popular, White hegemonic public sphere. This is what I infer to be Samuel DuBose's brother Aubrey's meaning when, in the media interview cited earlier, he credits a bodycam for making his brother's death at the hands of a police officer an event—in his words, a "story"—for both the news media and the District Attorney's office.

Sharpe links the infamous King beating video to another instance of initially aspirational and ultimately failed "imaging work" that precedes Ferguson and Black Lives Matter but belongs to the present era of social media–based public circulations of digital video of police brutality. This is the online sharing of cell phone camera footage of the shooting of Oscar Grant on an Oakland, California, BART train platform by White BART police officer Johannes Mehserle in the early morning hours of New Year's Day 2009. Grant was already restrained by another officer, face down on the platform with his hands behind his back, at the time of the shooting. The detained train car from which Grant and his friends had been evicted under suspicion of fighting was packed with fellow New Year's Eve revelers, many of whom recorded the event on their phones. Bystander video captured Grant's shooting in horrifically intimate detail, including the unmistakable sounds of Mehserle's gun.

The urgent task of public witnessing in these exemplary instances partly has to do with the videos being received and perceived as documentary evidence of state violence routinely directed at Black and Brown communities and rarely made visible—let alone legible as media spectacles—to either news media or juridical purview, in the form of criminal procedures or civil litigation. Moreover, their circulation performs a kind of perceptual metonymy linking anti-Black police violence to other biopolitical operations that disproportionality mark Black and Brown bodies in the United States for impairment and early death but which can elude recognition as *violence* because they are diffuse, suffered communally, and multifarious in their etiologies and thus unable to be depicted—visually or otherwise—within the singular parameters of an "event." These include pervasive malnutrition, addiction, lead-paint poisoning, lack of access to preventative medicine, and what public health researchers document as the mentally and physically disabling effects of racial discrimination and of the asymmetrical administration of criminal justice, including the vastly disproportionate incarceration of members of African American, Latinx, and Native communities.

And yet, the specific sense of urgency that attached to these videos in their original contexts of circulation was grounded in what was initially presumed to be their legal value as *evidence*. This is a set of referents these images no longer carry, given the acquittals of King's police assailants, and only an involuntary manslaughter conviction, in the case of Grant's murder. In the present, as they continue to circulate online, the King and Grant videos have acquired a different semiotic valence. It is this symbolism that Sharpe invokes when she makes their images proximate to the "imaging work" performed by the social media self-portraits of 2015's twenty-one murdered trans women. For, like the bodycam footage of DuBose's killing, or the surveillance camera video of the police shooting of eleven-year-old Tamir Rice, or civilian cellphone footage of the chokehold death of Staten Island father Eric Garner, they also now serve as emblems of an institutional failure to hold police accountable for the murder of unarmed African Americans.

The circulation of visual documentation of African Americans' violent mistreatment by police via social media, online news, the blogosphere, and televisual and print journalism becomes complicit in constructing the "racial profile" of the African American subject as one *a priori* targeted for the deaths these images depict. John L. Jackson describes graphic video documentation of police shootings of unarmed Black citizens as "snuff films," implying that the public's interest in such images is not just limited to its appetite for justice.[65] More than a literal description of death recorded on film, Jackson's analogy is grounded in the modalities of spectatorship and media production of the Internet, where, as El-Hadi notes, "Black death and dying are regularly recorded and uploaded, from Black persons killed by U.S. police, to . . . the drowned corpses of African migrants washing up on European shores."[66] The global Internet is, in this analysis, a continuum of Black suffering across the diaspora in which the victims of police shootings and fatal chokeholds comingle with the countless and undercounted deaths attributable to today's global refugee crisis.

A 2017 study funded by the city of Washington, DC, imputes a possible causal link between the adoption of body cameras and increased use-of-force and misconduct charges. This finding bears out the darkest implications of Jackson's suggestion that bodycams and bystander video might do very different kinds of political imaging and legal work than is putatively claimed on their behalf. It also raises a different set of questions than I initially set out to pose about the bodycam's constitutive role, as a media and legal technology, in an emergent structure of racialization I am calling post-panoptic. For if body cameras *encourage* rather than deter racial profiling and bias, it suggests that bodycams do not actually function, as their police and civilian proponents argue, and as their manufacturers claim, to surveil and, through disciplinary (including juridical) review, control the behavior of the people wearing them. Bodycams, considered from the perspective of this study's findings, refine the techniques of power and authority over the bodies they depict—that is, the racialized citizen-suspects who appear on their screens.

Academic research on the impact of the widespread adoption of the technology in the post-Ferguson era confirms this point. A recent report by criminologists at George Mason University, who surveyed seventy empirical studies on the effects of bodyworn cameras on policing in the United States, found that, when it comes to their role in the courts, body cameras are far more likely to be used to prosecute civilians than police officers.[67] And while the report also found some evidence that, at the statistical level, the numbers of complaints about police have gone down, a lower civilian rate of reporting on police might well be an effect of a pervasive erosion of civilians' sense of agency given the pervasive techno-surveillance involved in their encounters with police.

Due to the inordinate policing of African American and other communities of color, the routine presence of police bodycams in these communities naturalizes the objectification of certain bodies and what Sarah Ahmed has termed the "drastically unequal distribution of bodily vulnerabilities."[68] The metaphor of the interface that I introduced earlier is meant to illustrate the racial logic of the visual apparatus that extends through many aspects of modern policing (from bodycams to forensic evidence grids, from crime statistics mapping to algorithm-based predictive policing) and suggest how they function as a media logic. Currently, the opposing claims that surround the bodycam—that it both curbs and abets police violence, that it both undermines and preserves structural racism—point to its images' dual, contradictory function as part of the visual apparatus that reproduces state violence and also, through its images' mimetic kinship with civilian cellphone's anti-policing sousveillance videos, part of the visual dragnet that, in our hypermediated public sphere, has kept public focus magnetized to the state-sanctioned killing of Black women and men as one of the defining moral and political issues of our moment. What I have tried to bring to light vis-à-vis the visual and legal discourses and historical meanings that meet within the bodycam's surprisingly inscrutable lens is the bodycam's situation as a technique of racialization masquerading as race-neutral technology. Post-panopticism describes the media environment and underlying structures of social control that constitute the bodycam's platform, within which "race," "visuality," and "law" converge to form the interface that determines how lives *matter*, both literally and in the symbolic sense now so necessarily galvanizing for political discourse.

Notes

1. Bryce Clayton Newell, "Crossing Lenses: Policing's New Visibility and the Role of 'Smartphone Journalism' as a Form of Freedom-Preserving Reciprocal Surveillance," *University of Illinois Journal of Law, Technology, and Policity* 59 (2014).
2. Gilles Deleuze, "Postscript on the Societies of Control," *October* 59 (Winter 1992): 3–7.
3. See, especially, Neal Feigenson, "The Visual in Law: Some Problems for Legal Theory," *Law, Culture and the Humanities* 10, no. 1 (2014): 13–23; Jennifer L. Mnookin, "The Image of Truth: Photographic Evidence and the Power of Analogy," *Yale Journal of Law & Humanities* 10 (1998): 1.
4. Ira Torresi, "The Photographic Image: Truth or Sign?," *Law, Culture and Visual Studies* (2014): 126.
5. For a consideration of how this ideal of impersonality organizes testimony as a genre in the domain of law and other discursive contexts, see Lauren Berlant, "Trauma and Ineloquence," *Journal of Cultural Research* 5 (2001): 41–58.
6. Walter Benjamin, "The Work of Art in the Age of Mechanical Reproduction," *Visual Culture: Experiences in Visual Culture* (1936): 137–44.
7. Charles Sanders Peirce, *The Essential Peirce: Selected Philosophical Writings. Vol. 1*, ed. Nathan Houser and Christian Kloesel (Bloomington: Indiana University Press, 1992), 226.
8. In an influential essay that is foundational for studies of media technology and racial formation, Allan Sekula links the historical advent of photography to the rise of new

visual epistemologies of race, ethnicity, sexuality, and disability by which racialized hierarchies of human difference attained new legitimacy as institutionalized, professional knowledge about the human subject. See his "The Body and the Archive," *October* (1986): 3–64.

9. Eden Osucha, "The Whiteness of Privacy: Race, Media, Law," *Camera Obscura: Feminism, Culture, and Media Studies* 24, no. 1 (70) (2009): 67–107, quotation on 75.

10. Andrew John Goldsmith, "Policing's New Visibility," *British Journal of Criminology* 50 (2010): 914–34.

11. For studies of visual culture and media technologies that extend this project of analysis into later histories of media technology, to show how what Sekula terms photography's "instrumental realism" informs the racial reproductions of contemporary visual culture that comprise the broader discursive context shaping the representational capacities of images of police violence against African Americans from bodycams and other witnessing devices, see especially Wendy Hui Kyong Chun, "Introduction: Race and/as Technology; or, How to Do Things to Race," *Camera Obscura* (2009): 7–35; Ruha Benjamin, *Race after Technology: Abolitionist Tools for the New Jim Code* (Hoboken, NJ: John Wiley & Sons, 2019); Michael Boyce Gillespie, *Film Blackness: American Cinema and the Idea of Black Film* (Durham, NC: Duke University Press, 2016).

12. Relevant articles that suggest the standardization of this usage include Bradley A. Campbell, Justin Nix, and Edward R. Maguire, "Is the Number of Citizens Fatally Shot by Police Increasing in the post-Ferguson Era?," *Crime & Delinquency* 64, no. 3 (2018): 398–420; and Seth Wyatt Fallik, Ross Deuchar, and Vaughn J. Crichlow, "Body-worn Cameras in the Post-Ferguson Era: An Exploration of Law Enforcement Perspectives," *Journal of Police and Criminal Psychology* (2018): 1–11.

13. Carrie Myers Morrison, "Body Camera Obscura: The Semiotics of Police Video," *American Criminal Law Review* 54 (2017): 793.

14. Sekula, 5.

15. Sekula, 6–7.

16. Sekula, 10.

17. The implications of the findings for policing in general were limited by the study's design. Given the small size of the police force in Rialto, California, and the city's demographic, the department was not representative of policing across the country— larger urban areas, in particular. From a scientific perspective, the small sample size was limiting. Also, the short duration of the study raises questions about the durability of the body camera's putative positive effects.

18. Ben Brucato, "Policing Made Visible: Mobile Technologies and the Importance of Point of View," *Surveillance & Society* 13, no. 3/4 (2015): 460–62.

19. Brucato, 460.

20. Colby Itkowitz, "Michael Brown Petition Has 100,000 Signatures," *Washington Post*, August 20, 2014.

21. "Petition to the U.S. House of Representatives: Pass the Michael Brown, Jr. Law to Begin Equipping Police with Body Cameras," Change.org, accessed July 10, 2020, https://www.change.org/p/u-s-house-of-representatives-pass-the-michael-brown -jr-law-to-begin-equipping-police-with-body-cameras.

22. Her campaign platform reiterated the call for funding but avoided making cameras a mandatory requirement of federal law. See "Road to the White House 2016: Hillary Clinton Remarks at Columbia University," C-SPAN, April 29, 2015, https://www .c-span.org/video/?325657-1/hillary-clinton-remarks-criminal-justice-reform.

23. "Issues: Racial Justice," accessed February 12, 2020,https://berniesanders.com/ issues/racial-justice/.

24. Ben Jacobs, "Donald Trump Tells the Guardian Police Body Cameras 'Need Federal Funding,'" *The Guardian*, October 13, 2015, https://www.theguardian.com/us-news/2015/oct/13/donald-trump-police-body-cameras-federal-funding.

25. "End the War on Black People," *The Movement for Black Lives*, accessed February 12, 2020, https://m4bl.org/end-the-war-on-black-people/.

26. Howard Wasserman, "Moral Panics and Body Cameras," *Washington University Law Review* 92, no. 3 (2015): 831.

27. Debra Thompson, "An Exoneration of Black Rage," *South Atlantic Quarterly* 116, no. 3 (2017): 458.

28. Thompson, 460.

29. Thompson, 459.

30. See, for example, http://www.chicagotribune.com/bluesky/technology/chi-body-cameras-hidden-costs-20150206-story.html; http://www.freep.com/story/news/local/michigan/2016/06/06/police-body-cameras-high-costs/85356518/.

31. See "Police Body-Worn Camera Legislation Tracker," *Urban Institute*, last modified October 19, 2018, http://apps.urban.org/features/body-camera-update/; "Body-Worn Camera Laws Database," *National Conference of State Legislatures*, last modified February 28, 2018, http://www.ncsl.org/research/civil-and-criminal-justice/body-worn-cameras-interactive-graphic.

32. Ian Cummings, "$16,000 to See Sarasota Police Body Camera Video," *Sarasota Herald-Tribune*, March 2, 2015, http://www.heraldtribune.com/news/20150302/16000-to-see-sarasota-police-body-camera-video.

33. The Ray Tensing trial and its aftermath is the focus of one chapter of a 2017 four-part *New York Times* video series, available for streaming as a single documentary on the newspaper's online video channel, *Times Documentaries*. The series explores police body cameras' social and legal implications and effects on policing in America. In the *Times* film, Audrey DuBose's poster is centrally showcased and can be observed throughout the crowds gathered in support of her family at public rallies in Cincinnati. Brent McDonald and Hilary Bachelder, "The Rise of Body Cameras," *New York Times*, January 13, 2017, https://www.nytimes.com/video/us/100000004859799/the-rise-of-body-cameras.html.

34. Morrison, 793.

35. Morrison, 800.

36. Associated Press, "Some Police Departments Halt Use of Body Cameras, Citing Costs," *New Haven Register*, September 10, 2016.

37. Nick Swartsell, "Samuel Dubose's Shooting Shows How Important Police Body Cameras Can Be," *Vice*, July 31, 2015, https://www.vice.com/en_us/article/5gj98a/cops-lies-and-videotape-in-cincinnati-731.

38. Steve Martinot, "On the Epidemic of Police Killings," *Social Justice* 39, no. 4 (2014): 53.

39. Dexter Thomas, "Perspective: Cincinnati Body-cam Images Made Us Players in a Video Game," *Los Angeles Times*, July 30, 2015, http://www.latimes.com/nation/nationnow/la-na-nn-dubose-shooting-footage-real-life-video-game-20150729-story.html.

40. Morrison, 834.

41. Morrison, 817.

42. Morrison, 813.

43. Olevia Boykin, Christopher Desir, and Jed Rubenfeld, "A Better Standard for the Use of Deadly Force," *New York Times*, January 1, 2016, www.nytimes.com/2016/01/01/opinion/a-better-standard-for-the-use-of-deadly-force.html, cited in Thompson, 471.

44. Morrison, 820.

45. Saidiya V. Hartman, *Scenes of Subjection: Terror, Slavery, and Self-making in Nineteenth-century America* (Oxford: Oxford University Press, 1997), 22.

46. Hartman, 57–58.

47. Hartman, 38.

48. Christina Sharpe, *In the Wake: On Blackness and Being* (Durham, NC: Duke University Press, 2016).

49. Nicholas Mirzoeff, *The Right to Look: A Counterhistory of Visuality* (Durham, NC: Duke University Press, 2011).

50. Mirzeoff, 167.

51. Richard Dyer, *White: Essays on Race and Culture* (New York: Routledge 1999), 23.

52. Lauren Berlant, *The Queen of America Goes to Washington City: Essays on Sex and Citizenship* (Durham, NC: Duke University Press, 1997), 176.

53. Jaime Hellman, "No Clear Picture on How Many People Are Killed by Police," *Al Jazeera America*, May 28, 2015, http://america.aljazeera.com/watch/shows/Ali-Velshi-On-Target/articles/2015/5/28/No-data-on-how-many-killed-by-police-in-America.html.

54. See http://gawker.com/what-ive-learned-from-two-years-collecting-data-on-poli-1625472836.

55. "Fatal Force," *Washington Post*, accessed November 22, 2020, https://www.washington post.com/graphics/investigations/police-shootings-database/.

56. John Sullivan, Reis Thebault, Julie Tate, and Jennifer Jenkins, "Number of Fatal Shootings by Police Is Nearly Identical to Last Year," *Washington Post*, July 1, 2017, https://www.washingtonpost.com/investigations/number-of-fatal-shootings-by-police-is-near ly-identical-to-last-year/2017/07/01/98726cc6–5b5f-11e7–9fc6-c7ef4bc58d13_story.html.

57. See, for example, Juliet Hooker, "Black Protest/White Grievance: On the Problem of White Political Imaginations Not Shaped by Loss," *South Atlantic Quarterly* 116, no. 3 (2017): 483–504; Shatema Threadcraft, "North American Necropolitics and Gender: On# BlackLivesMatter and Black Femicide," *South Atlantic Quarterly* 116, no. 3 (2017): 553–79; Ben Brucato, "Fabricating the Color Line in a White Democracy: From Slave Catchers to Petty Sovereigns," *Theoria* 61, no. 141 (2014): 30–54; Kashif Jerome Powell, "Making# BlackLivesMatter: Michael Brown, Eric Garner, and the Specters of Black Life—Toward a Hauntology of Blackness," *Cultural Studies? Critical Methodologies* 16, no. 3 (2016): 253–60; Kenneth Lawson, "Police Shooting of Black Men and Implicit Racial Bias: Can't We All Just Get Along," *University of Hawai'i Law. Review* 37 (2015): 339; Charles R. Lawrence III, "The Fire This Time: Black Lives Matter, Abolitionist Pedagogy and the Law," *Journal of Legal Education* 65 (2015): 381.

58. Leigh Raiford, *Imprisoned in a Luminous Glare: Photography and the African American Freedom Struggle* (Chapel Hill: University of North Carolina Press, 2011).

59. Leon F. Litwack, "Hellhounds," in *Without Sanctuary: Lynching Photography in America*, ed. James Allen, Hilton Als, John Lewis, and Leon F. Litwack (Santa Fe, NM: Twin Palms, 2000), 13. See also Amy Louise Wood, "Lynching Photography and the 'Black Beast Rapist' in the Southern White Masculine Imagination," in *Masculinity: Bodies, Movies, Culture*, ed. Peter Lehman (New York: Routledge, 2001), 205, 198–99.

60. Litwack, 10–11.

61. Wood, 205.

62. See Alexandra Juhasz, "How Do I (Not) Look? Live Feed Video and Viral Black Death," *Jstor Daily* 20 (2016), https://daily.jstor.org/how-do-i-not-look/.

63. Sharpe, *In the Wake*, 117.

64. Sharpe (emphasis mine).

65. John L. Jackson, "Lights, Camera, Police Action!," *Public Culture* 28, no. 1 (2016): 4.

66. Nehal El-Hadi, "Death Undone," *The New Inquiry* May 2 (2017).

67. Cynthia Lum, Megan Stoltz, Christopher S. Koper, and J. Amber Scherer, "Research on Body-worn Cameras: What We Know, What We Need to Know," *Criminology & Public Policy* 18, no. 1 (2019): 93–118.

68. Sarah Ahmed, "Selfcare as Warfare," feministkilljoys, August 25, 2014, https://feministkilljoys.com/2014/08/25/selfcare-as-warfare/.

CHAPTER 5

Visualizing the Surveillance Archive

Critical Art and the Dangers of Transparency

TORIN MONAHAN

> Archives—as records—wield power over the shape and direction of historical scholarship, collective memory, and national identity, over how we know ourselves as individuals, groups, and societies.
>
> > —Joan M. Schwartz and Terry Cook, "Archives, Records, and Power: The Making of Modern Memory"

> But if we grant that symbolic systems are social products that contribute to making the world, that they do not simply mirror social relations but help *constitute* them, then one can, within limits, transform the world by transforming its representation.
>
> > —Loïc J.D. Wacquant, "Toward a Social Praxeology: The Structure and Logic of Bourdieu's Sociology"

As public concerns about state surveillance ebb and flow, with people buffeted from one revelation to the next, demands for transparency assert a powerful organizing force on responses. Transparency often serves as the initial impulse motivating disclosures, as a way of shedding light on hidden surveillance programs in order to activate change, and as an objective in the aftermath of public revelations, as actors seek to discover more details about the programs and practices in question. The predictable valorization of transparency should be interrogated, though, especially given that historical precedent provides ample reason to doubt that transparency will automatically lead to accountability or substantial changes in law or policy.[1]

Perhaps ironically, the affective comfort provided by the discourse of transparency can be traced to the dominance of modern scientific rationality, which privileges decision-making based on surveillance, quantification, and data analysis.[2] Desire for a knowable and controllable world is coded into the DNA of contemporary institutions as part of the legacy of the scientific revolution and its twentieth-century emphasis on efficiency and rationality. In other words, institutional quests for transparency and control lead to many of the surveillance abuses that the public seeks to rein in through similar strategies, beginning with making them visible. Transparency efforts usually concentrate on publicizing existing compendiums of state or corporate activities of monitoring and intervening in the affairs of others. That said, discourses of transparency tend to ignore the politics of archival categories and rationalities upon which transparency depends.

Critical artworks may help us perceive some of the dangers and limitations of transparency, even if many of them fall victim to a similar trap of failing to problematize the normative valence of archival designs. Thus, this chapter investigates a number of critical surveillance artworks that grapple with transparency. What the analyzed artworks share is a general critique of ubiquitous, undemocratic surveillance operations that invisibly and unequally structure people's lives. These are the elements that artists problematize through their creations. Some of them seek to reveal the materiality of state surveillance infrastructures, such as secret military installations and satellites; others demonstrate the objectifying effects of routine corporate surveillance systems like Google Street View; and still others take advantage of social media platforms to intentionally oversaturate viewers with personal images as a way to highlight the irrationality of government programs targeting suspected terrorists. They each confront institutional surveillance by constructing *counter-archives* of visual material. If transparency is the goal, then creating counter-archives is a way to transform surveillance programs into objects of scrutiny and critique. It is also a way to contest the legitimacy of institutional archives (of the National Security Agency, Google, etc.) by underscoring their partiality and incompleteness. To grasp the potential of counter-archiving efforts, however, one must also reckon with the general politics of archives in order to assess the extent to which archival practices shape and constrain resistance efforts.

Politics of the Archive

Archives are technologies of power. Just as with the historical record, what is included and how the items are organized create truth claims in ways that are insulated from critique by nature of the implied neutrality of the technical apparatus—classification schemes, cabinets and shelves, algorithms, digital storage media, and so on. Limitations to participation in archive construction or access reinforce exclusions that have political effects, as the legitimacy of cultural experience or knowledge is shaped by the degrees to which one can contribute to and mobilize archival materials.[3] Further, as Allan Sekula persuasively argues, archives perform a pedagogical function of normalizing social hierarchies and communicating one's place in those hierarchies; for instance, through exposure to circulated photographs of suspected criminals or honorific figures, one is trained to accept social hierarchies, recognize visual signifiers of value, and aspire for self-improvement to climb the social ladder.[4]

In his influential writings on institutions and the human sciences, Michel Foucault set the stage for inquiry into the constitutive force of archives.[5] He notes how archives do not merely *reflect* changes in ways of seeing and being but function as instruments of administrative power to *produce* those changes. The archive operates as a discursive system that determines what can be thought or said in any particular historical moment.[6] Jacques Derrida extends some of these observations to stress the importance of archives in articulating forms of political power: "There is no political power without control of the archive, if not memory. Effective democratization can always be measured by this essential criterion: the participation in and access to the archive, its constitution, and its interpretation."[7]

Because archival structures set the parameters for what can be included, the design of archives also embodies and advances political agendas that are often disguised by idealistic notions of the archive as a repository of materials for advancing the public good.[8] Indeed, notwithstanding contemporary views of archives as safeguarding "public memory," the earliest archives were those dedicated to the establishment of "royal memory" and sovereign power for administrative control of populations and territories.[9] What has been called the "imperial archive" produced instrumental and sometimes grossly inaccurate representations of colonial topographies, resources, and cultures, which were then imposed upon colonial subjects for purposes of resource extraction and population control.[10] Colonial archives, and their legacies, facilitated political division, oppression, and persecution, with the examples of apartheid in South Africa or genocide in Rwanda being just a few of the most notorious cases where arbitrary colonial classification and record systems set the foundation for racial violence.[11] The "totalitarian archives" of Nazi Germany and communist East Germany, among those in other countries, further illustrate the capacity of thorough, rationally produced and administered archives to subtend dehumanizing systems of oppression and killing.[12] Importantly, as Michael Lynch argues, the decisions made in building an archive serve an interpretive function that channels future uses in particular directions, so there is no such thing as a "raw" archive.[13] It is vital, therefore, to confront the politics of those initial and ongoing decisions, as well as the politics of archiving imperatives more broadly.

Even the concept of a "public archive," which is a form often romanticized as fostering equality and democracy,[14] was grounded from its inception in practices of surveillance and social order. For example, as richly detailed by Patrick Joyce,[15] the advent of the free public library in mid-nineteenth-century Britain was designed to cultivate a liberal citizen who embraced a self-help ethos that depended on close monitoring of one's deficiencies and one's progress at correcting them in the quest for self-improvement. In order to assist in this project, libraries collected data on the social characteristics of local populations, including their unique customs and dialects, to generate a public archive that could be used to diagnose social problems and direct cultural improvement projects, for instance, in partnership with educational institutions. The physical design of libraries began to change at this time as well, moving away from common tables where scholars could engage in discussion and resource sharing to individual stations in an open panoptic space, which afforded individuation of responsibility (for checking out books, paying fines, etc.) and observation by others, effectively transforming citizens into a "public police" to protect the archive from theft.[16] Because of the public archive's role in the formation of liberal subjects, one could view it as "a political expression of the nation state itself."[17]

Archival practices betray additional affinities with the surveillance of everyday life. The organizing principles of "archival reason," as theorized by sociologist Thomas Osborne, include publicity (making information public), singularity (focusing on detail), and mundanity (concentrating on the everyday).[18] Archival reason either guides surveillance activities or sets the conditions for them, particularly by providing a professional framework and impulse for the collection and cataloging of fine-grained, oftentimes mundane data in the interest of publics, even if those publics are sometimes narrowly delimited or exclusive. Importantly, this is a significant shift in what types of materials are assumed to be valuable, as "mundane" details reside not just in community or personal practices but in all institutional contexts as well. Osborne asserts:

> So it is not just a question of a romantic focus upon the powerless. What is at stake here, in fact, is a distinctive way of making visible the question of power itself. If royal memory was a memory of the sovereign and great acts, then archival memory in its modern forms is a memory—even when it focuses on the great and the powerful themselves—of everyday detail; a style of memory that contains within itself the assumption that the everyday is a particularly revealing level on which to pose the question of memory.[19]

Archival reason, described in these terms, clearly undergirds contemporary surveillance practices, where the systematic amassing, classifying, and processing of mundane data is regarded as essential to the functioning of responsible modern subjects and organizations.

The law, frequently viewed as the legitimate response to surveillance abuses, also depends upon and reproduces archival logics. Abstractly, the law similarly adopts archival reason to concentrate attention on evidence, mundane detail, and precedent in its continuous reconstruction of a particular notion of justice, the liberal subject, and the public good. What we might think of as the legal archive would include more than simply legal doctrines and other trappings of legal thought but also—following Sekula's insights about the social ramifications of metaphoric archives[20]—the role of the law in the ongoing constitution of the social. As Rosemary Coombe explains:

> The law's greatest cultural impact may be felt where it is least evident . . . the law is working not only when it is encountered in its most authoritative spaces, but also when it is consciously and unconsciously apprehended. The moral economies created in the shadows of law, the threats of legal action made as well as those that are carried out, people's everyday fears and anxieties about the law, are all loci where the law is doing cultural work.[21]

Beyond legal structures alone, then, the figure of the law operates through discourse and practice to shape subjectivities, with differential effects based on one's social position and biography.[22] This does not imply that the law is monolithic or totalizing, but rather that it signifies an assemblage of power relations whose contours and intensities vary radically depending on the sites of its materialization.

In that the law is perpetually in flux and negotiated through practice,[23] it provides a context for surveillance-based archiving and counter-archiving. For instance, government spying programs often push upon or cross legal thresholds in their capacious archiving of communications data, whereas artist-activist-scholars sometimes operate on the margins of legality (e.g., trespassing or revealing state secrets) to compile and circulate counter-archives of those state programs and sites. The law offers rich zones of contestation, in part because it is so mutable and fragile, contrary to popular perceptions of it as being static and stable. Nonetheless, because of "law's primary allegiance to liberal political theory,"[24] it participates in the reproduction of capitalist ideological orders and contributes to processes of individuation,[25] frequently in ways that amplify social inequalities and intersectional forms of oppression.[26] Thus, to the extent that countersurveillance efforts embrace archival reason and appeal to the law, their radical potential may be blunted.

Visualizations of Critical Surveillance Art

In a time of information overload and constant crises, visualizations acquire prophetic overtones. As Orit Halpern explains, visualization is a way of bringing that which is absent into view for purposes of speculation and action.[27] The implied rationality of institutional surveillance and archiving practices, which strive to make the world knowable and governable, depends upon forms of visualization. This carries over into big-data discourses as the pursuit of total knowledge through technological efforts to map and represent unintuitive and unpredictable relations among disparate data.[28] That said, recent trends in data visualization prioritize aesthetics (colors, patterns, texture) over information clarity in their attempts to communicate complexity; they also stress the sublime, mystical overtones of "beautiful data," which perform simultaneously in both affective and logical registers.[29] The surveillance visualizations of government agencies and technology companies now extend well beyond the video walls and closed-circuit television screens made vernacular by popular science-fiction films and television shows to include dynamic, three-dimensional topographical schematics, network maps, sentiment analysis diagrams, and more. There is performative force in the beauty of these objects, fueling their uptake and symbolically communicating omniscience even as their designers offer caveats and acknowledge limitations, mere whispers drowned out by the symphony of visual support.

Most critical surveillance art projects capitalize upon the visual force of surveillance representations while also attempting to destabilize institutional surveillance demands. They do so by making hidden surveillance programs perceptible through counter-archiving and visualization projects as well as by critiquing the violence of surveillance operations that reduce human complexity to manageable data elements. In other words, these artworks, like government or commercial visualizations, also navigate between aesthetics and clarity, but with different goals. They aspire not to total transparency of the entire surveillance enterprise but instead to provocative traces, glimpses, and clues that function metonymically to direct critical attention to the reductive and controlling logics of institutional surveillance. In this respect, they approximate "diagrammatic thinking," a mode of engagement that sketches problem spaces and the conditions for imagining new social configurations.[30]

While artists have different motivations and aspirations, it is important to interrogate the ways in which their critical surveillance projects enact archival reason. We can ask: What are the assumptions behind the construction of counter-archives? How do such counter-archives perform, and what are their politics? As a starting point, these projects generate hidden transcripts of state and corporate surveillance practices and construct oppositional visualizations, suggesting the recalibration of power relations between individuals and institutions. As such, they appear to place great stock in the political efficacy of transparency. They also depend upon archival modalities that have historically functioned in the service of the state, even—or especially—in the implementation of the liberal archive that individuates and disciplines subjects. Thus, my analysis of these art projects probes the ramifications of responding to hidden systems of visualization with open ones, of constructing *aesthetic archives of disclosure* that poach upon institutional surveillance infrastructures.

Traces of the State

There is something elusive about state surveillance operations and the infrastructures upon which they depend. Beyond official conditions of secrecy, which limit exposure and awareness, contemporary surveillance resists investigation by means of its opaque materiality and bureaucratic banality. To visualize such surveillance requires more than pulling back the veil and showing it for what it is. Instead, it necessitates the cultivation of new ways of seeing.[31]

Through investigative research and photographic methods, the work of geographer and artist Trevor Paglen illustrates the potential of artistic efforts to document the mundane materiality of technological infrastructures in ways that invite viewers to reflect on the hidden worlds of state surveillance. In photographing military drones, undersea fiber-optic cables, and secret military bases, just to name a few targets, Paglen transposes the tenets of archival reason into immersive surveillant landscapes. In museum exhibitions, for instance, he presents hazy panoramic images of massive white radomes punctuating an otherwise idyllic English countryside in Cornwall (see Figure 4),[32] which is a site that serves as a satellite surveillance outpost for the UK's Government Communications Headquarters and the US National Security Agency.[33] Another image, this time a somber underwater landscape off the coast of Miami, depicts a blurry snake of fiber optic cable hugging the sandy floor as it extends from the lower right and disappears into the murk at the center of the frame. This cable is one of many allegedly tapped by the NSA to gather global telecommunications data for its surveillance programs, which was a practice revealed by Edward Snowden in 2013.[34]

The motivation behind these photographs is to eradicate secrecy, which, according to Paglen, "nourishes the worst excesses of power."[35] It is clear from Paglen's projects that state secrecy is his primary concern, and the excesses that alarm him have to do not only with unchecked data collection, but with violence against other human beings, as with drone strikes or rendition and torture of suspected terrorists, and with the rise of a shadow military and intelligence complex.[36] That said, while seeking transparency through the fabrication of a counter-archive, Paglen recognizes

Figure 4. Trevor Paglen, "National Security Agency Surveillance Base, Bude, Cornwall, UK" (2014). Courtesy of the artist.

the insufficiency and perhaps even unwitting complicity of these efforts: "History shows that revelation and change can often work in tandem, but revelation, in and of itself, accomplishes little."[37] Moreover, Paglen notes, "When covert actions and classified programs become public, their revelation is often used to legitimize their profoundly troubling purposes, to sculpt the state in their own image."[38] In other words, when confronted with unwelcome disclosures, such as with illegal NSA spying programs or CIA torture practices, the state moves not to scale back, necessarily, but to retroactively legalize those activities and grant immunity to those who participated in them.[39]

What makes Paglen's work compelling is that the counter-archives he builds function simultaneously as evidence and as art. As forms of evidence, his photographs of specific sites may have little political or legal currency,[40] but the overall performance of constructing a counter-archive of state secrets signifies a direct challenge, communicating that one can infiltrate and document these hidden worlds, even if, it must be acknowledged, Paglen reproduces tropes of heroic masculinity (e.g., as the intrepid adventurer putting himself in danger) to generate photographs for his archive. As art forms, Paglen's photographs resonate in entirely different ways. While the motivation for their production might be transparency, as visual artifacts they resist decoding and assimilation in that it is often not immediately clear what his blurry images depict, how they were made, or what message is intended. In journalist Jonah Weiner's interpretation: "Blurriness serves both an aesthetic and an 'allegorical' function. It makes his images more arresting while providing a metaphor for the difficulty of uncovering the truth in an era when so much government activity is covert."[41] More subtly, Paglen explains that his works are "a way of organizing your attention" upon that which, like the sublime, moves you through its magnitude and incomprehensibility.[42] By focusing perception on the materialities of state surveillance and violence, he tries to make visible what the state is at pains to erase through techniques of archival control (e.g., classified or redacted records, censored maps). In the process, his work sets the conditions for the transformation of subjectivities, for viewers to develop *a feel* for the systems of which they are a part and for which they share responsibility.[43]

Apart from Paglen's work, complementary projects by other artists add perspective on the politics of this mode of counter-archiving. Tomas van Houtryve's project "Blue Sky Days," for example, presents an archive of photographs generated by the artist using a personal drone flying over US sites where the Federal Aviation Administration had granted permission for state drone use.⁴⁴ By making images of everyday events—such as weddings or children's birthday parties—from the perspective of a drone, van Houtryve lays a frame of violence over the prosaic, highlighting similarity among people and encouraging a sense of shared vulnerability. The counter-archive is not a "real" one of US drones or government-produced images; instead, it operates on the level of the imaginary to engender disquiet through forced juxtaposition of what is seen and what is unseen but known, namely the "collateral damage" of US drone strikes in other countries. In so doing, this public counter-archive draws attention to and stands in for the absent public archive of drone operations in non-US contexts.

Josh Begley's "Plain Sight: The Visual Vernacular of NYPD Surveillance" takes a different tack of appropriating and repackaging police images of Muslim neighborhoods, businesses, and places of worship.⁴⁵ These surveillance images were produced by the New York Police Department's Demographics Unit as part of its counter-terrorism operations following the attacks of 9/11. As Begley explains:

> Plain Sight is an attempt to catalog the banality and violence of the visual culture of the NYPD's secret spying unit. . . . Plain Sight re-presents that which has already been made public, and draws attention to the often mundane, innocuous, and indiscriminate nature of the information collected. Through this work, viewers are provided with different entry points through which they can see surveillance and examine this historical archive in a new light.⁴⁶

Using mostly images of buildings and storefronts, the artist forms a circular collage evoking a human eye, with pastel pinks and blues around the perimeter, as the iris, and darker, black and white tones in the center, as the pupil. This molded presentation of NYPD's visual archive rejects static or neutral interpretations of the images and instead draws attention to a very specific act of looking. It emphasizes the illegal profiling of certain populations in the absence of any evidence of wrongdoing. Whereas inclusion in the police archive was an act of symbolic violence, Begley's response offers a way of destabilizing the symbolic regime established by such police surveillance by subjecting it to scrutiny.⁴⁷

The works reviewed in this section each trace state-surveillance practices, whether through documentation (Paglen), simulation (van Houtryve), or compilation (Begley). The artists assemble counter-archives of photographic material to both visualize and contest secret state operations. At the same time, these counter-archives do not aspire toward totalization. Rather, through blurry representations, suggested absent archives, and repackaged appropriations of visual evidence, they acknowledge the multiplicity of interpretive possibilities and in the process subvert the state's hold on symbolic authority. Thus, whereas the act of manufacturing counter-archives tacitly valorizes transparency, with included documents intended to serve as a resource for other forms of political engagement, the artists concurrently reject realism. Their archives are objects to think with but not pathways to the truth. In this respect, the logics of archiving are fissured within these works: archival reason prevails, with its focus on collecting, organizing, and publicizing mundane detail, while at the same time representations of the truth are problematized. Where these efforts congeal most with previous articulations of archivization is in their implicit construction of liberal subjects. The artworks position viewers as agents responsible for their own political transformation, through the edification initiated by the artists, and the transformation of state practices, through individual or collective action.[48] The objectives might be oppositional to state surveillance, but the archival mechanisms by which they are pursued harmonize well with the impulses that motivate modern state actions to begin with.

Ghosts in the Archive

Although much surveillance scholarship concentrates on the state and its governance regimes, corporate data systems now permeate and structure most aspects of everyday life in industrialized countries. The archives and algorithms of just about all industries (financial, healthcare, real estate, transportation, and so on) shape life chances in largely invisible—and often unequal—ways, dictating what resources and services one has access to based upon assessments of one's relative risk or value.[49] Industries obviously partner with government agencies as well, whether voluntarily or through legal requirements, to share data pertaining to possible national security threats or to act as surrogates for the state to do things like freeze bank accounts or prevent individuals from flying.[50] Government agencies also provide a lucrative market for security products and services, allowing for the easy flow of knowledge, personnel, and equipment across public and private sectors.[51] So while clear distinctions cannot hold between the practices of public and private entities, artists and scholars have recognized that voracious data collection by private companies must be included in critiques of contemporary surveillance.

Data behemoths like Google offer fruitful material for artistic probing of the corporate archive. Artist Paolo Cirio's "Street Ghosts" project, for instance, repurposes images from Google Street View (the application that delivers photographic representations of streetscapes) to draw attention to and problematize Google's visual archive of individuals in public space.[52] For this project, Cirio locates images of people in Google Street View and paints life-size representations of those blurred figures in the exact location in the urban landscape where the original photos were made. One work, for example, superimposes onto a rusted and graffitied wall an incongruous figure of a balding man in a long-sleeve pink shirt and light blue jeans (see Figure 5). Overgrown weeds rise up around the man's lower legs, obscuring his feet, as he tilts slightly to the side with what appears to be a spade in one hand and a plant in the other. He seems to be gardening in the exact spot where the weeds now overtake his past efforts.[53]

This piece performs on several levels. It is a glimpse into the Google archive and a signification, through a singular figure, of the vast totality of personal images contained within it. By reproducing images of people in space, Cirio removes Google images from their corporate frame and imposes an artistic one that harkens back to, but does not recapture, the original. In the process, he calls attention to the violence of stripping data from context and challenges the unsanctioned production of personal images—images that become Google's property upon their creation. On another level, "Street Ghosts" fabricates a counter-archive that is etched into the built world. It comprises two-dimensional surface illustrations of what came before, communicating the incompleteness of all representation, including those of our many "data doubles" in institutions' digital archives.[54] Unlike many of those data doubles, Cirio's figures also stage their own temporality and degradation, visibly eroding over time as analog instruments of memory and mortality.[55] They readily abandon their fleeting gesture toward the real, becoming palimpsests or backdrops for other urban articulations, other

Figure 5. Paolo Cirio, "Street Ghosts" (2014). Courtesy of the artist.

embodiments of meaning in space. Thus, contrary to Google's archive, even though this counter-archive points back to the moment of image creation, it is decidedly nonindexical. It offers intentionally ephemeral and ghostlike representations of representations, undermining the implied realism of the Google archive by demonstrating the necessary but arbitrary function of framing in establishing meaning or truth.

A different project by artist Andrew Hammerand moves critique beyond corporate archives themselves to the pervasive surveillance infrastructures that enable them. In a piece called "The New Town," Hammerand accesses an unsecure webcam mounted on a cellphone tower in a generic Midwestern suburban community in the United States to build an image archive of the quotidian activities of residents.[56] There are pixilated and fragmented images of what appear to be young teenagers in a park: a girl hugs someone from behind as another leans into a cartwheel, and another appears between the other three with only her upper body in the frame; apparently the video stream had not refreshed in time to show her lower half. Another image shows the silhouette of a woman sitting at an outdoor table as a brown-haired woman in a white T-shirt struts by, with her head and shoulders appearing two strides before the rest of her, separated from her body by another digital glitch. In another frame, one can vaguely make out two people meeting on a pathway through the grass and perhaps shaking hands or exchanging an item. For Hammerand, the project provides a way to "reflect on the expansive use of surveillance technologies, the increasing loss of privacy, and the heightened sense of anxiety and vulnerability that are part of American life in the early 21st century."[57]

In spite of the artist's stated intentions, themes of anxiety and vulnerability are not self-evident in "The New Town." One could interpret this work as a commentary on the obliviousness or nonchalance of individuals to surveillance exposure, on the lack of anxiety about being inserted in archives without one's awareness or consent. Approached as a counter-archival performance, however, the fractured images and ambiguity of action destabilize conventional understandings of images. The many visual defects violently segment bodies, making everyday life tainted and illegible. The screenshots cannot offer objective representations of people or events because context always exceeds the frame and the frame itself is flawed. Without being organized in archives and placed into discourse, meaning appears ambiguous and incomplete, which is a finding that highlights that the power of archives and discourses rest in their ability to reduce and deny the multiplicity of meaning always inherent in representations. Moreover,

"The New Town" draws attention to the vulnerabilities of corporate surveillance infrastructures, which can be hijacked to spy on others or produce alternative images.

The question remains whether the production of such counter-archives undermines or solidifies visual economies of image generation and exchange. Corporate control over archives is a way of ensuring profit, as can be seen with any digital enclosure that safeguards intellectual property, but the contribution to and proliferation of image archives by others also becomes a resource for appropriation by corporate and state actors.[58] This can be witnessed with the disparate actions of Facebook using vast repositories of user images to finetune their facial-recognition algorithms for tailored advertising;[59] Flickr making a rapacious claim to be able to sell and profit from the images uploaded to its platform;[60] or police agencies scouring social media sites as part of their investigations of suspects.[61] Seen from this perspective, projects like Hammerand's "The New Town" capitalize upon existing surveillance infrastructure to generate new images for art, but in the process mimic the very archival practices that they critique: contributing to image streams with no apparent regard for the privacy of their subjects and no clear targets for intervention. Projects like those of Paolo Cirio, which reclaim digital images and insert them into physical space, also compromise subjects' privacy—in mirroring Google's act of doing so—but may be better equipped to render archival politics visible. Such works contest specific corporate monopolies on archiving, namely Google's, while also destabilizing discourses of visual realism, ultimately affirming the value of transient embodied moments within specific local contexts.

Archival Oversaturation

In the contemporary conjuncture, most institutions are devoted to and constituted by archival logics. They endeavor to transform all elements of their domains of operation into data to be harnessed for decision-making, regardless of whether they are in the public, private, or non-profit sectors. Systems for the efficient production and management of data become essential within this context. Thus, tapping into the social-media streams of individuals or capturing the "digital exhaust" of their online and offline activities (e.g., web searches, purchases, entertainment choices, travel locations) offer some of the most attractive ways to build valuable digital archives through automated or crowdsourced data generation. As people are entrained to participate in systems of data production, they become compliant surveillance subjects whose labor generates value for others, often unbeknownst to them.[62] However, the very mechanisms that allow for and encourage participation by users in data creation also afford opportunities for artistic interventions that illustrate the insidious *illogic* of supposedly rational systems.

The work of artist Hasan Elahi plays upon this apparent contradiction. Following the attacks of 9/11, Elahi—who at the time was an assistant professor at San Jose State University—was wrongly placed on an FBI watch list and detained for interrogation in 2002 at Detroit Metropolitan Airport.[63] He was investigated and repeatedly questioned for months before finally having his name cleared, at which point he was advised to continue to inform the FBI of his travel activity.[64] His response was to inundate the FBI with an overabundance of photographic documentation about his daily whereabouts, purchases, toilets used, food eaten, and more. He created a public website (trackingtransience.net) to disseminate this photographic evidence, as a performance and critique of state demands for transparency from victims of unwarranted suspicion and profiling. By 2014, Elahi had produced a counter-archive of more than 70,000 images,[65] each of them another datum presumably conveying his innocence, pushed into FBI data streams as counterevidence to stifle and mock whatever spurious material led to his targeting in the first place.

Elahi's counter-archive oversaturates viewers, confounding attempts to make definitive interpretations about what is revealed or what really matters.[66] This strategy hinges on the premise that "more is less,"[67] that privacy can be maintained by drowning others in one's personal data without giving them a key for deciphering the many elements depicted or registering their relative importance. More than that, copious disclosure masks selective nondisclosure. Elahi explains:

> I still stand behind the idea that you protect your privacy by giving it up. . . .
> For now, we're still in a transition between analog and digital, and for as long
> as we're in this state of flux, we'll develop a more sophisticated understanding
> of the consequences of living under constant surveillance. For now, at least, we
> still have control over what information we put forth publicly. Being mindful
> of how we do that feels like a good first step toward retaining control.[68]

Oversaturation, therefore, has a double meaning in the context of Elahi's work. It is an analytic device communicating, on one hand, that there is too much visual material to process and, on the other hand, that just as photographs lose their detail and take on an unnatural appearance with an increase in color intensity, so too do surveillance subjects lose definition with archival oversaturation. This, at least, is the primary conceit of Elahi's Tracking Transience website.

Elahi's massive counter-archive, while functioning as a critical art project in itself, has served as a resource for numerous spin-off projects as well. For instance, the work titled "Prism," named after one of the NSA's Internet spying programs, is a large 10 × 36 foot poster of vertical, multicolored bars overlaid on a black-and-white background image of the roof of the NSA building (see Figure 6).[69] The colored bars evoke those displayed on televisions during Emergency Broadcast System tests, which originated during the Cold War period in the United States as part of the civil defense program to manage public communications in the event of nuclear attack.[70] Upon closer inspection, each of the bars is composed of thousands of photographs from Elahi's self-surveillance archive: meals, urinals, shopping centers, airports, and so on. This work tacitly asserts that within surveillance societies, national states of emergency metamorphose into scrutiny of the minute details of people's lives. Elahi compares the contemporary situation to that of people under the Stasi secret police in East Germany prior to the fall of the Soviet Union; the difference is that he is the one voluntarily sharing intimate details, rather than that information coming from a secret spy network of informants.[71] "Prism," then, acts as a critique of the totalitarian tendencies of modern states in their response to perceived insecurities.

Figure 6. Hasan Elahi, "Prism" (2015). Courtesy of the artist.

These tendencies may be grounded in fears with some empirical basis, but they are animated by the insatiable archival appetite of security institutions like the NSA, whose members want desperately to absorb and process *all* communications data throughout the world.[72]

If total knowledge is the goal of security agencies, Elahi challenges the depth and precision of conclusions drawn predominantly from surface representations. For example, another project of his titled "Stay VI.0" is a detailed print of many of the dozens of beds Elahi has slept in during his travels (see Figure 7).[73] Not a single bed is well made. The covers and sheets are bunched and twisted, while pillows are askew and frequently stacked, all seen from the perspective of someone standing at the foot of the beds. As with all of Elahi's work, photographs of people, including of himself, are conspicuously absent, so beyond the fact of an unmade bed, viewers can ascertain very little about the artist from these pictures. We do not know if he slept well, if he slept alone, if or what he read or watched there, or if stayed in the room all night. Each bed stands in for information that it cannot convey, and adding more photographs of similar beds does not increase depth of awareness; it only solidifies the pattern of partial, insignificant revelation. As art scholar Sven Spieker relates: "Archives do not record experience so much as its absence; they mark the point where an experience is missing from its proper place, and what is returned to us in an archive may well be something we never possessed in the first place."[74] Elahi's work suggests,

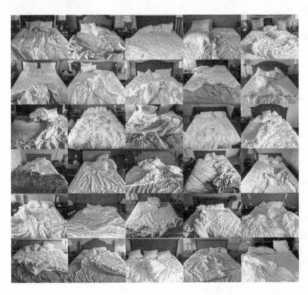

Figure 7. Hasan Elahi, "Stay VI.0" (2011). Courtesy of the artist.

by extension, that law enforcement or security agency archives, in their expansiveness, produce a false sense of definitive knowledge of subjects when in fact they are best at reifying their protocols of classification and imposing them upon individuals.

Elahi draws upon and simultaneously deconstructs photography's historical aura of realism and authenticity. By nature of their verisimilitude and original co-presence with that which they represent, photographs appear to have an indexical relationship to the physical world, but this indexical quality has always been a fabrication.[75] As a medium, photographs are necessarily shaped by the technical valences and constraints of photographic devices, as well as by the many other cultural factors that go into staging and processing photographic media.[76] The very language used to talk about photographs obscures this process of construction: the vernacular privileges hunting metaphors of "shooting" photos or "capturing" scenes, when in fact all images are "made."[77] Additionally, photography is always embedded in discourses and institutional practices that work diligently to maintain the appearance of accuracy and singularity of meaning, thereby silencing interpretive multiplicity, especially with respect to law enforcement's use of photographs as evidence. John Tagg explains:

> The function of photography as a means of surveillance, record, and evidence was the result of a more or less violent struggle . . . to hold in place certain discursive conditions. It depended on a machinery of capture that sought to curtail the productivity of photographic meanings, exhaust their legibility, and make the camera its own, as an instrument of a new disciplinary power.[78]

Elahi pushes back on this disciplinary power by appropriating the trope of photographic evidence and claiming the right to self-document, multiplying representations of his interfacing with objects until meaning collapses under the symbolic weight of the archive.

At the same time, archival reason prevails. Elahi contests the legitimacy of law enforcement archives by producing a parallel archive to perform innocence. While the primary audience has long ceased to be institutional agents, if it ever was, these works depend upon an oppositional framing that maintains the centrality of the hidden archives of the surveillance state. These state archives are the ones that matter, the ones that are consequential in determining the conditions of one's existence. The appeal to self-transparency, even if self-evidence is highly selective or intentionally misleading, redounds to valorization of photographic realism. It asks viewers to hold these archives side by side—the visible one of quotidian scenes and the hidden one of incriminating evidence—and conclude that the artist is innocent and the state illegitimate. In this sense, Elahi's projects assume an underlying rational framework of disinterested and impartial assessment of competing archives, ignoring the ways in which power asymmetries infuse not only the application of archival material but their formation as well. Rather than protecting privacy by giving it up, the impulse to produce and circulate photographic material reaffirms belief in the efficacy of archives to materialize just and democratic outcomes despite ample evidence to the contrary.

Conclusion

Through the assembly of counter-archives in the pursuit of transparency, artists provocatively frame state and corporate surveillance systems as objects of critique. They endeavor to make institutional archives and infra-structures transparent so that the public might challenge existing institu-tional monopolies on symbolic authority. Key to this is not only circulating artist-generated images but also multiplying interpretive possibilities for all visual evidence. The focus here has not been on the *intentions* of artists per se but on the cultural work performed by artistic counter-archives. The art projects emphasize mediated traces and transience, alongside negotiated data representations, all in contradistinction to the presumed fixity and opacity of institutional surveillance archives. Thus, for most of the works discussed in this chapter, meaning is approached as being indeterminate and contingent, shaped by discursive regimes, technological protocols, and personal experience. Counter-archive images, in this context, do not seek to establish definitive representations of reality from the artists' per-spectives so much as to draw attention to the interpretive dimensions of all visual materials, including those curated by institutional actors. That said, counter-archives *do* perform as oppositional structures. Although their individual elements might splinter connections to the real, taken as a whole counter-archives assert competing truth claims to combat those of institutions, and thereby recuperate the real in the final instance, calling upon viewing subjects to choose a version of the truth and act upon it.

Ultimately, counter-archives operate as variations of the liberal archive. They interpellate viewers as subjects charged with negotiating their rela-tionship with the state (e.g., to safeguard privacy, prove innocence, demand accountability) and therefore affirm the ultimate authority of the state in arbitrating those efforts. The implied appeal to the state to reform illegal surveillance practices, for instance, accedes the legitimacy of the police and the law, when those institutions are complicit in the infractions perpetrated by the state and by corporate entities.[79] It is a call upon the state to police itself when it is best designed to police subjects. Additionally, especially with regard to the state's construction of abject bodies or suspect identities, exceptions abound, permitting extra-legal surveillance, interrogation, and detention, which are practices that are often normalized and legalized after the fact, making appeals to the law impotent.[80] As Anthony Giddens perceptively observed, "Intensified surveillance and totalitarian tendencies are intimately linked,"[81] so one should not be surprised by these institutional developments or the role of the liberal archive in supporting them.

Thus, the archival mechanisms deployed by artists harmonize well with the rationalities that animate surveillance abuses in the first place. They embrace transparency and apply archival reason to construct representations of discrete, mundane events and circulate them with the goal of catalyzing specific interventions. Rather than question visualization imperatives, the artists instead take issue with certain kinds of institutional visualizations, which they seek to amend with the production of more—but different— images. While the presumed audience for these art works may be the broad public, many of the visual media produced are elite and exclusionary: expensive art books, museum installations, inscrutable websites. Therefore, the democratic potential of these works is likely attenuated by the conservative conventions of art presentation. Finally, many of the technological platforms and protocols utilized by these artists are themselves surveillant, and institutional actors readily capture the data produced by them. For instance, when disseminated online, these counter-archives enter into larger data streams such that they are absorbed into institutional archives, processed for content and meta-data analysis, and linked or fed to targeted viewers, thus becoming content for a larger system of surveillance capitalism.[82] Whether through the production of counter-archives, mechanisms of legal discovery, or other means, transparency can never be a sufficient response to institutional surveillance as long as such imbricated power asymmetries persist. The pursuit of it invariably circles back to surveillance-based visuality as a privileged form of knowledge and governance.

Notes

1. Ben Brucato, "The New Transparency: Police Violence in the Context of Ubiquitous Surveillance," *Media and Communication* 3, no. 3 (2015): 39–55; Byung-Chul Han, *The Transparency Society* (Stanford, CA: Stanford University Press, 2015); Kregg Hetherington, *Guerrilla Auditors: The Politics of Transparency in Neoliberal Paraguay* (Durham, NC: Duke

University Press, 2011); Torin Monahan, "Surveillance and Terrorism," in *Routledge Handbook of Surveillance Studies*, ed. Kirstie Ball, Kevin D. Haggerty, and David Lyon (London: Routledge, 2012).

2. Anthony Giddens, *The Consequences of Modernity* (Stanford, Calif.: Stanford University Press, 1990); Theodore M. Porter, *Trust in Numbers: The Pursuit of Objectivity in Science and Public Life* (Princeton, NJ: Princeton University Press, 1995); James C. Scott, *Seeing Like a State: How Certain Schemes to Improve the Human Condition Have Failed* (New Haven, CT: Yale University Press, 1998).

3. Hetherington, *Guerrilla Auditors*.

4. Allan Sekula, "The Body and the Archive," *October* 39 (1986): 3–64.

5. Michel Foucault, *The Order of Things: An Archaeology of the Human Sciences* (New York: Vintage Books, 1970); Foucault, *The Archaeology of Knowledge and The Discourse on Language*, trans. A. M. Sheridan Smith (New York: Pantheon Books, 1972).

6. Foucault, *The Archaeology of Knowledge and the Discourse on Language*, 129.

7. Jacques Derrida, "Archive Fever: A Freudian Impression," *Diacritics* 25, no. 2 (1995): 9–63: 11, cited in Marlene Manoff, "Theories of the Archive from Across the Disciplines," *Portal: Libraries and the Academy* 4, no. 1 (2004): 9–25.

8. Manoff, "Theories of the Archive from Across the Disciplines."

9. Jacques Le Goff, *History and Memory* (New York: Columbia University Press, 1992); Thomas Osborne, "The Ordinariness of the Archive," *History of the Human Sciences* 12, no. 2 (1999): 51–64; see also Scott, *Seeing Like a State: How Certain Schemes to Improve the Human Condition Have Failed*.

10. It is important to remember the central role of the geographical and anthropological sciences in constructing these imperial maps and archives.

   James Duncan, "Complicity and Resistance in the Colonial Archive: Some Issues of Method and Theory in Historical Geography," *Historical Geography* 27 (1999): 119–28; George E. Marcus, "The Once and Future Ethnographic Archive," *History of the Human Sciences* 11, no. 4 (1998): 49–64; David H. Price, *Cold War Anthropology: The CIA, the Pentagon, and the Growth of Dual Use Anthropology* (Durham, NC: Duke University Press, 2016); Thomas Richards, *The Imperial Archive: Knowledge and the Fantasy of Empire* (New York: Verso, 1993). More than simply saying that the social sciences have been complicit in egregious state practices, as some form of cautionary tale, it would be better to face the fact that the social sciences and their theoretical frameworks emerged in conjunction with—and within—the modern state, in the service of its military and capitalistic interests, and to question the extent to which that legacy continues to shape scholarly inquiry and assessment of its value today. See also Gayatri Chakravorty Spivak, "The Rani of Sirmur: An Essay in Reading the Archives," *History and Theory* 24, no. 3 (1985): 247–72.

11. Geoffrey C. Bowker and Susan Leigh Star, *Sorting Things Out: Classification and Its Consequences* (Cambridge, MA: MIT Press, 1999); David Lyon, "Identification Practices: State Formation, Crime Control, Colonialism and War," in *Technologies of InSecurity: The Surveillance of Everyday Life*, ed. Katja Franko Aas, Helene Oppen Gundhus, and Heidi Mork Lomell (New York: Routledge-Cavendish, 2009).

12. John O. Koehler, *Stasi: The Untold Story Of The East German Secret Police* (Boulder, CO: Westview Press, 1999); Lyon, "Identification Practices: State Formation, Crime Control, Colonialism and War"; Osborne, "The Ordinariness of the Archive"; see also Alfred W. McCoy, *Policing America's Empire: The United States, the Philippines, and the Rise of the Surveillance State* (Madison: University of Wisconsin Press, 2009).

13. Michael Lynch, "Archives in Formation: Privileged Spaces, Popular Archives and Paper Trails," *History of the Human Sciences* 12, no. 2 (1999): 65–87.

14. Take as one authoritative example, a passage from the International Council on Archives' *Universal Declaration on Archives*: "[Archives] play an essential role in the

development of societies by safeguarding and contributing to individual and community memory. Open access to archives enriches our knowledge of human society, promotes democracy, protects citizens' rights and enhances the quality of life." Anne J. Gilliland and Sue McKemmish, "The Role of Participatory Archives in Furthering Human Rights, Reconciliation and Recovery," *Atlanti: Review for Modern Archival Theory and Practice* 24 (2014): 78–88.

15. Patrick Joyce, "The Politics of the Liberal Archive," *History of the Human Sciences* 12, no. 2 (1999): 35–49.

16. Joyce.

17. Joyce, 36

18. Osborne, "The Ordinariness of the Archive."

19. Osborne, 59.

20. Sekula, "The Body and the Archive."

21. Rosemary Coombe, "Is There a Cultural Studies of Law?" in *A Companion to Cultural Studies*, ed. Toby Miller (Malden, MA: Blackwell, 2001), 57

22. Mindie Lazarus-Black and Susan F. Hirsch, eds., *Contested States: Law, Hegemony, and Resistance* (New York: Routledge, 1994); Sally Engle Merry, "Rights Talk and the Experience of Law: Implementing Women's Human Rights to Protection from Violence," *Human Rights Quarterly* 25, no. 2 (2003): 343–81.

23. Kitty Calavita, *Invitation to Law and Society: An Introduction to the Study of Real Law* (Chicago: University of Chicago Press, 2010).

24. Julie E. Cohen, "Studying Law Studying Surveillance," *Surveillance & Society* 13, no. 1 (2015): 91–101, quote at 92.

25. Louis Althusser, *On the Reproduction of Capitalism: Ideology and Ideological State Apparatuses*, trans. G. M. Goshgarian (London: Verso, 2014).

26. Kimberle Crenshaw, "Demarginalizing the Intersection of Race and Sex: A Black Feminist Critique of Antidiscrimination Doctrine, Feminist Theory and Antiracist Politics," *University of Chicago Legal Forum* (1989): 139–67; John Gilliom, *Overseers of the Poor: Surveillance, Resistance, and the Limits of Privacy* (Chicago: University of Chicago Press, 2001); Stuart Hall et al., *Policing the Crisis: Mugging, the State and Law and Order*, 2nd ed. (New York: Palgrave Macmillan, 2013).

27. Orit Halpern, *Beautiful Data: A History of Vision and Reason since 1945* (Durham, NC: Duke University Press, 2014), 21.

28. Mark Andrejevic, "Surveillance in the Big Data Era," in *Emerging Pervasive Information and Communication Technologies (PICT)*, ed. Kenneth D. Pimple (New York: Springer, 2014).

29. Halpern, *Beautiful Data*; Anthony McCosker and Rowan Wilken, "Rethinking 'Big Data' as Visual Knowledge: The Sublime and the Diagrammatic in Data Visualisation," *Visual Studies* 29, no. 2 (2014): 155–64.

30. Gilles Deleuze, "The Diagram," in *The Deleuze Reader*, ed. Constantin V. Boundas (New York: Columbia University Press, 1993); McCosker and Wilken, "Rethinking 'Big Data' as Visual Knowledge: The Sublime and the Diagrammatic in Data Visualisation."

31. Torin Monahan, "Ways of Being Seen: Surveillance Art and the Interpellation of Viewing Subjects," *Cultural Studies* 32, no. 4 (2018): 560–81.

32. Radomes are microwave radar units encased in a protective dome structure to insulate them from the elements and from scrutiny. David Wood, "Territoriality and Identity at RAF Menwith Hill," in *Architectures: Modernism and After*, ed. Andrew Ballantyne (Malden, MA: Blackwell, 2004).

33. Tim Sohn, "Trevor Paglen Plumbs the Internet," *New Yorker*, September 22, 2015, http://www.newyorker.com/tech/elements/trevor-paglen-plumbs-the-internet-at-metro-pictures-gallery, accessed June 11, 2017.

34. Glenn Greenwald, *No Place to Hide: Edward Snowden, the NSA, and the U.S. Surveillance State* (New York: Metropolitan Books, 2014); Sohn, "Trevor Paglen Plumbs the Internet."

35. Jonah Weiner, "Prying Eyes," *New Yorker*, October 22, 2012, http://www.newyorker
    .com/magazine/2012/10/22/prying-eyes, accessed June 11, 2017.
36. Trevor Paglen, *Blank Spots on the Map: The Dark Geography of the Pentagon's Secret
    World* (New York: Dutton, 2009).
37. Paglen, 280.
38. Paglen, 277.
39. John Gilliom and Torin Monahan, *SuperVision: An Introduction to the Surveillance
    Society* (Chicago: University of Chicago Press, 2013); Paglen, *Blank Spots on the Map*.
40. One important counterpoint was with Paglen's photographs of a "black site" detention
    facility in Afghanistan. When the US government tried to prevent Majid Khan—a US
    legal resident who was subjected to extraordinary rendition and tortured by the CIA—
    from obtaining legal counsel because he might reveal secret information about his
    treatment, Paglen's published photographs were cited by the Center for Constitutional
    Rights as legal evidence that the site in question was not, in fact, secret. Ben Davis, *9.5
    Theses on Art and Class* (Chicago: Haymarket Books, 2013); Weiner, "Prying Eyes."
41. Weiner, "Prying Eyes."
42. Weiner, 56, 60
43. Then again, according to some observers of audiences at Paglen's exhibits, the "art-
    seeking class" in attendance seems more interested in the social event of consuming
    art than the politics: "Those who lingered were drawn to the video installation's lush
    images and pulsing soundscape, although many glanced away periodically at their
    phones, texting and e-mailing and posting photos, as though they hadn't quite got
    the message." Sohn, "Trevor Paglen Plumbs the Internet."
44. Tomas van Houtryve, "Blue Sky Days," *Open Society Foundations* 2014, https://www.
    opensocietyfoundations.org/moving-walls/22/blue-sky-days, accessed June 11, 2017.
45. Josh Begley, "Plain Sight: The Visual Vernacular of NYPD Surveillance," *Open
    Society Foundations* 2014, https://www.opensocietyfoundations.org/moving-walls/22
    /plain-sight-visual-vernacular-nypd-surveillance, accessed June 11, 2017.
46. Begley.
47. Begley's collage also highlights the *volume* of the police archive in a way that individual
    images could not connote.
48. This responsibilizing motif can be seen in full force at one of Trevor Paglen's exhibits
    with the implementation of a Tor node for anonymous web browsing. Kashmir Hill,
    "Art That Shows Us What Mass Surveillance Actually Looks Like," *Fusion*, September
    20, 2015, http://fusion.net/story/199240/trevor-paglen-art-shows-what-mass-sur
    veillance-looks-like/, accessed December 26, 2016. Museumgoers can *choose* the
    anonymous network for their Internet activity and thereby make a small, individual-
    ized contribution to the protection of privacy in society. The fact that an institution
    (the museum) provides such a node is perhaps more consequential, as a symbolic
    statement that could be replicated by other institutions. That said, the emphasis is on
    threats to privacy, as opposed to other social problems facilitated by surveillance, aligns
    well, of course, with the individualizing slant of the liberal archive. Torin Monahan,
    "Regulating Belonging: Surveillance, Inequality, and the Cultural Production of
    Abjection," *Journal of Cultural Economy* 10, no. 2 (2017): 191–206.
49. Solon Barocas and Andrew D. Selbst, "Big Data's Disparate Impact," *California
    Law Review* 104, no. 3 (2016): 671–732; Oscar H. Gandy Jr., *Coming to Terms with
    Chance: Engaging Rational Discrimination and Cumulative Disadvantage* (Burlington,
    VT: Ashgate, 2009); Tarleton Gillespie, "The Relevance of Algorithms," in *Media
    Technologies: Essays on Communication, Materiality, and Society*, ed. Tarleton Gillespie,
    Pablo J. Boczkowski, and Kirsten A. Foot (Cambridge, MA: MIT Press, 2014); Frank

Pasquale, *The Black Box Society: The Secret Algorithms That Control Money and Information* (Cambridge, MA: Harvard University Press, 2015).

50. Anthony Amicelle, "Towards a 'New' Political Anatomy of Financial Surveillance," *Security Dialogue* 42, no. 2 (2011): 161–78; Louise Amoore, *The Politics of Possibility: Risk and Security Beyond Probability* (Durham, NC: Duke University Press, 2013); Kirstie Ball et al., *The Private Security State? Surveillance, Consumer Data and the War on Terror* (Copenhagen: Copenhagen Business School Press, 2015).

51. Philip Boyle and Kevin D. Haggerty, "Spectacular Security: Mega-Events and the Security Complex," *International Political Sociology* 3, no. 3 (2009): 257–74; Ben Hayes, "The Surveillance-Industrial Complex," in *Routledge Handbook of Surveillance Studies*, ed. Kirstie Ball, Kevin D. Haggerty, and David Lyon (London: Routledge, 2012); Torin Monahan, "The Future of Security? Surveillance Operations at Homeland Security Fusion Centers," *Social Justice* 37, no. 2–3 (2011): 84–98. Through the law, the state also protects the ecology for corporate data practices and profits, as seen, for instance, with the refuge created for copyright law in online environments. Hector Postigo, *The Digital Rights Movement: The Role of Technology in Subverting Digital Copyright* (Cambridge, MA: MIT Press, 2012); Julie E. Cohen, *Configuring the Networked Self: Law, Code, and the Play of Everyday Practice* (New Haven, CT: Yale University Press, 2012); Tarleton Gillespie, *Wired Shut: Copyright and the Shape of Digital Culture* (Cambridge, MA: MIT Press, 2007).

52. Paolo Cirio, "Street Ghosts," *Open Society Foundations*, 2014, https://www.opensocietyfoundations.org/moving-walls/22/street-ghosts, accessed June 12, 2017. Chris Ingraham and Allison Rowland have written about complementary artistic efforts to stage unreal *tableaux vivants* (living images) for Google Street View cameras, including events like emergency childbirths and crime scenes, in order to resist the reductive frames of Google archives by "performing imperceptibility." Chris Ingraham and Allison Rowland, "Performing Imperceptibility: Google Street View and the Tableau Vivant," *Surveillance & Society* 14, no. 2 (2016): 211–26.

53. By returning to Google's original image, the man does not seem to be gardening, after all, but instead performing some kind of maintenance work around a fire hydrant. Paolo Cirio, "Album Archive: Street Ghosts—Public," 2013, https://get.google.com/albumarchive/101628793830161569774/album/AF1QipP66ok3UiNs8t4wovmvBOSJgHTfspSw7IOgLZNV/AF1QipPQVd9svwHBqIC2RZTv5R-gobERYJFjdm2ckOiY, accessed June 12, 2017. The image is true to the Google original, but easy to misread—or read differently—as the context fluctuates. The discrepancies highlight the vital role of framing and the necessary interpretive dimension of reading photographs. Meaning is always fluid, never fixed.

54. *Data doubles* are abstract and partial representations of individual bodies in data, usually acted upon instrumentally by institutions. Kevin D. Haggerty and Richard V. Ericson, "The Surveillant Assemblage," *British Journal of Sociology* 51, no. 4 (2000): 605–22.

55. This is not to say that the digital does not degrade too, as can be noted with frantic efforts to translate digital content before media disintegrate or protocols become obsolete. Jean-François Blanchette, "A Material History of Bits," *Journal of the Association for Information Science and Technology* 62, no. 6 (2011): 1042–57.

56. Andrew Hammerand, "The New Town," *Open Society Foundations* 2014, https://www.opensocietyfoundations.org/moving-walls/22/new-town, accessed June 12, 2017.

57. Hammerand.

58. Mark Andrejevic, "Estranged Free Labor," in *Digital Labor: The Internet as Playground and Factory*, ed. Trebor Scholz (New York: Routledge, 2013); Nicole S. Cohen, "The Valorization of Surveillance: Towards a Political Economy of Facebook," *Democratic Communiqué* 22, no. 1 (2008): 5–22.

59. Richard A. Spinello, "Privacy and Social Networking Technology," *International Review of Information Ethics* 16, no. 12 (2011): 41–46.

60. Stuart Dredge, "Flickr Takes Flak for Selling Creative Commons Photos as Wall-art Prints," *The Guardian*, December 2, 2014, https://www.theguardian.com/technology/2014/dec/02/flickr-creative-commons-photos-wall-art, accessed June 12, 2017.

61. Daniel Trottier, "Police and User-led Investigations on Social Media," *Journal of Law, Information and Science* 23, no. 1 (2014): 75–96.

62. Mark Andrejevic, *iSpy: Surveillance and Power in the Interactive Era* (Lawrence: University Press of Kansas, 2007); Shoshana Zuboff, "Big other: Surveillance Capitalism and the Prospects of an Information Civilization," *Journal of Information Technology* 30, no. 1 (2015): 75–89; Valerie Steeves, "Hide and Seek: Surveillance of Young People on the Internet," in *Routledge Handbook of Surveillance Studies*, ed. Kirstie Ball, Kevin D. Haggerty, and David Lyon (London: Routledge, 2012).

63. Clive Thompson, "The Visible Man: An FBI Target Puts His Whole Life Online," *Wired.com*, May 22, 2007, https://www.wired.com/2007/05/ps-transparency/, accessed June 12, 2017.

64. Hasan Elahi, "FBI, here I am!," *TEDGlobal*, July 2011 https://www.ted.com/talks/hasan_elahi, accessed June 12, 2017,.

65. Elahi, "Thousand Little Brothers," *Open Society Foundations* 2014, https://www.opensocietyfoundations.org/moving-walls/22/thousand-little-brothers, accessed December 26, 2016.

66. Simon Hogue, "Performing, Translating, Fashioning: Spectatorship in the Surveillant World," *Surveillance & Society* 14, no. 2 (2016): 168–83; Elise Morrison, *Discipline and Desire: Surveillance Technologies in Performance* (Ann Arbor: University of Michigan Press, 2016).

67. Rachel Hall, Torin Monahan, and Joshua Reeves, "Editorial: Surveillance and Performance," *Surveillance & Society* 14, no. 2 (2016): 154–67.

68. Hasan Elahi, "I Share Everything. Or Do I?," *Ideas.TED.com*, July 1, 2014, http://ideas.ted.com/i-share-everything-or-do-i/, accessed June 12, 2017.

69. Elahi, "Prism," 2015, http://elahi.umd.edu/elahi_prism.php, accessed June 12, 2017; Tim Smith, "Hasan Elahi Addresses Surveillance in Multilayered Exhibit at C. Grimaldis Gallery," *Baltimore Sun*, May 24, 2016, http://www.baltimoresun.com/entertainment/arts/artsmash/bs-ae-arts-story-0527-20160526-story.html, accessed June 12, 2017.

70. Elahi, "Thousand Little Brothers."

71. Elahi.

72. Greenwald, *No Place to Hide.*

73. Hasan Elahi, "Stay v1.0," 2011, https://lareviewofbooks.org/article/art-mattersnow-12-writers-on-20-years-of-art-orit-gat-on-2006-and-the-rise-of-youtube/, accessed November 9, 2020.

74. Sven Spieker, *The Big Archive: Art from Bureaucracy* (Cambridge, MA: MIT Press, 2008), 3

75. Marita Sturken and Lisa Cartwright, *Practices of Looking: An Introduction to Visual Culture*, 2nd ed. (Oxford: Oxford University Press, 2009); Mary Ann Doane, "Indexicality: Trace and Sign: Introduction," *differences* 18, no. 1 (2007): 1–6.

76. Jonathan Crary, "Techniques of the Observer," *October* 45 (1988): 3–35; Kelly Gates, "The Cultural Labor of Surveillance: Video Forensics, Computational Objectivity, and the Production of Visual Evidence," *Social Semiotics* 23, no. 2 (2013): 242–60.

77. Gillian Rose, *Visual Methodologies: An Introduction to Researching with Visual Materials*, 3rd ed. (Thousand Oaks, CA: Sage, 2012).

78. John Tagg, *The Disciplinary Frame: Photographic Truths and the Capture of Meaning* (Minneapolis: University of Minnesota Press, 2009), xxviii.

79. Brad Evans, *Liberal Terror* (Malden, MA: Polity, 2013); Mark Neocleous, *The Fabrication of Social Order: A Critical Theory of Police Power* (Sterling, VA: Pluto Press, 2000); Tyler Wall, "Ordinary Emergency: Drones, Police, and Geographies of Legal Terror," *Antipode* 48, no. 4 (2016): 1122–39.

80. Torin Monahan, *Surveillance in the Time of Insecurity* (New Brunswick, NJ: Rutgers University Press, 2010).

81. Anthony Giddens, *The Nation-State and Violence: Volume 2 of A Contemporary Critique of Historical Materialism* (Berkeley: University of California Press, 1987), 341.

82. Zuboff, "Big Other: Surveillance Capitalism and the Prospects of an Information Civilization."

CHAPTER 6

Becoming Invisible

Privacy and the Value of Anonymity

BENJAMIN J. GOOLD

Introduction

> We live in a surveillance society. It is pointless to talk about sur-
> veillance society in the future tense. In all the rich countries of
> the world everyday life is suffused with surveillance encounters,
> not merely from dawn to dusk but 24/7. Some encounters ob-
> trude into the routine, like when we get a ticket for running a red
> light when no one was around but the camera. But the majority
> are now just part of the fabric of daily life.
>
> —Kirstie Ball et al., *A Report on the Surveillance Society*

> A desire for privacy does not imply shameful secrets. . . .
> Without anonymity in discourse, free speech is impossible, and
> hence also democracy. The right to speak the truth to power
> does not shield the speaker from the consequences of doing so;
> only comparable power or anonymity can do that.
>
> —Nick Harkaway, *The Blind Giant: How to Survive in
> the Digital Age*

The law renders us visible in a multitude of ways. It assigns us labels
within the justice system ("accused," "offender," "witness," "victim") and
confers various types of status ("citizen," "resident," "refugee," "spouse,"
"minor") that mediate many of our day-to-day interactions with the state
and its agencies. The law also structures how the state "sees" us and how
information about us is collected, stored, processed, used, and shared.
How the law frames ideas of consent and disclosure, how it circumscribes

boundaries between the public and the private, and how it understands the connection between information and identity all help to shape the contours of the relationship between the individual and the state and, more fundamentally, to determine the limits of the state's ability to know and define us.

For the most part, we experience this process of being made visible through law as routine. The idea that the state can (and should) be allowed to gather information about who we are and how we live our lives does not seem especially controversial to most people. We accept that the state needs to know certain things about us in order to make decisions about how to use resources, provide services, and ensure good government. We willingly (if sometimes begrudgingly) fill out our tax returns, complete census forms, and answer security questions at the airport. We understand that in order to do certain things—like drive a car or travel internationally—we must be willing to be identified by the state. Increasingly, we also recognize and accept that our online activities may be the subject of scrutiny, as news reports draw attention to the ability of the police and other law enforcement agencies to monitor Internet traffic, search our phones, and read our email (ostensibly in the name of crime prevention and security). We not only surrender basic biographical information in exchange for being recognized but also submit to being defined—by gender, race, country of origin, and occupation—according to fixed categories that have been predetermined by the state.

In part, we do all this because it simply makes our lives easier. But underlying the acquiescence that comes from convenience is something deeper: we are willing to provide information and allow ourselves to be known because we trust the state. In modern liberal democracies, we generally assume that the state is benign (if not always efficient or competent) and that the information it collects from us will be used for our benefit. In recent years, however, such customary confidence in the state has faced serious challenge, and it has become more common, even in ostensibly liberal democracies like the United States, for ordinary citizens to question whether the state can in fact know too much about us.[1] More specifically, concerns about the growing use of sophisticated forms of electronic surveillance in the United States and countries throughout Europe have led to public debates about whether the law places adequate restrictions on the surveillance powers of the state.[2]

Even before Edward Snowden's 2013 disclosures regarding the operation of various global surveillance programs—orchestrated by the US National Security Agency (NSA) in conjunction with partner agencies in

Australia, Canada, New Zealand, and the United Kingdom[3]—academics, privacy advocates, and NGOs like the ACLU, Human Rights Watch, and the Electronic Privacy Information Centre (EPIC) had expressed concerns about increasing levels of surveillance by the police and other law enforcement agencies.[4] Indeed, such concerns were being voiced even before 9/11 and the subsequent introduction of legislation like the PATRIOT Act in the United States and the Anti-Terrorism Crime and Security Act 2001 in the United Kingdom. Writing in the early 1970s, the sociologist James Rule raised concerns about the potential dangers of mass electronic surveillance by the state:

> The growth of mass surveillance and control, then, seems somehow bound up with the changing structures of modern societies. But just what is the nature of this association? And what new forms of social organization and control do further developments along these lines promise? Does the continued extension of mass surveillance promise the advent of more and more oppressive forms of social control? Is the association between rigid mass surveillance and authoritarian rule accidental or inevitable? . . . Does the continued growth of mass surveillance draw us relentlessly into a world like Orwell's?[5]

While social theorists and activists have long held such concerns about state surveillance, what is notable about the last decade is that these concerns have begun to be shared by the public at large. According to a 2017 survey conducted by Reuters in the United States, 75 percent of adults would not allow law enforcement agencies to monitor their Internet activity in order to combat domestic terrorism.[6] The same survey also found that 37 percent of adults were of the view that "U.S. intelligence agencies are conducting too much surveillance on American citizens." Similarly, public opinion surveys conducted in the UK and Europe after the Snowden revelations have consistently shown that the majority of those under the age of sixty regard current levels of state surveillance of digital communications as excessive and an infringement of the right to privacy.[7] Taken together, the results of such surveys suggest that the general public has become alive to the fact that we may have, in the words of former UK Information Commissioner Richard Thomas, "sleepwalked" into a surveillance society.[8]

In the United Kingdom, the most visible manifestation of the establishment of a surveillance society has been the rapid proliferation of public area closed-circuit television (CCTV) cameras. By the late 2000s, it was estimated that there were five million video surveillance cameras operating in Britain, with a substantial number of them—upward of 59,000—being operated by local authorities (often in conjunction with the police) and overlooking public spaces such as city streets, shopping districts, and parks.[9] Although the use of such cameras has not been as widespread either in the United States or across Europe, the last twenty years has nonetheless been marked by a significant expansion in the use of such technology by state agencies around the world. In New York city alone, it is estimated that there are over 13,000 CCTV cameras providing continuous video surveillance of streets, parks, subways, and public housing.[10] In France and Germany, public area surveillance cameras were virtually unheard of until the late 1990s, but now the use of CCTV in major urban centers like Paris and Berlin is increasingly common.

In addition to the growing proliferation of CCTV cameras, other less visible forms of state-sponsored surveillance have also been on the rise. As the technology of surveillance becomes cheaper and more readily accessible, routine monitoring of domestic communications and Internet use by the police and security services—ostensibly to gather information on threats from terrorism and intelligence about organized crime—has become increasingly widespread.[11] While the Snowden disclosures drew attention to the fact that states frequently engage in intrusive forms of communications surveillance beyond their territorial boundaries, they also exposed just how much agencies like the NSA also engage in domestic monitoring. As both the ACLU and EPIC have noted, it is clear from sources like the Snowden documents that US security services and law enforcement agencies have the capacity to engage in "dragnet surveillance," collecting vast amounts of personal information and contextual data—from phone records, online searches, and social media—that enable them to develop highly detailed "pattern-of-life" pictures of individuals. Moreover, advances in surveillance technology have made the supposed boundary between international and domestic surveillance so blurred that the restrictions set out in legislation like the *FISA Amendments Act* 2008 (FAA) are essentially meaningless. As noted by the ACLU:

> The FISA Amendments Act of 2008 (FAA) gives the NSA almost unchecked power to monitor Americans' international phone calls, text messages, and emails—under the guise of targeting foreigners abroad. The ACLU has long warned that one provision of the statute, Section 702, would be used to eavesdrop on Americans' private communications. . . . [Recent disclosures] also show that an unknown number of purely domestic communications are monitored, that the rules that supposedly protect Americans' privacy are weak and riddled with exceptions, and that virtually every email that goes into or out of the United States is scanned for suspicious keywords.[12]

At the same time as states have been increasing their domestic and international surveillance capacities, many governments have also sought to limit the ability of individuals to conceal information about themselves or to protect their communications from state scrutiny. As will be discussed in the next section, even in those jurisdictions where privacy has the status of a legal right, it is rarely placed on the same footing as rights such as freedom of expression or freedom of religion. Instead, privacy rights are typically subject to various qualifications—most notably in the name of crime prevention and national security—and often poorly understood by legislators and the courts.[13] Privacy, in this regard, is best understood as a weak right, and as such, it typically provides only limited protection against the surveillance power of the state.

Similarly, the idea that individuals should be able to communicate anonymously is, in the words of Michael Froomkin, under "sustained legal and practical attack."[14] In addition to detailing the various ways in which governments have overtly sought to undermine anonymity—including the imposition of various online identification requirements and increasingly strict data retention laws—Froomkin has noted that intelligence services like the NSA have actively fought against anonymity by undermining national standards for cryptography and using FISA requests to demand user data from companies like Google, Facebook, Microsoft, Twitter, and Yahoo. In short, individuals have been caught in a pincer movement—between an ever-expanding network of state-sponsored surveillance systems on the one side and concerted efforts to weaken individual privacy and deny the possibility of anonymity on the other.

Although a great deal has been written about all of these developments—by journalists, activists, lawyers, and academics—many of the concerns have centered around the implications of state surveillance for individuals, in particular what is lost as a result of the assault on privacy. Framed in terms of a tension between privacy and security (or some other legitimate state aim), we are asked to consider just how much personal information we are willing to surrender in order to be made safer or more secure, with the assumption that there is some line in the sand that most people will eventually be willing to draw in terms of incursions on their privacy. While many privacy advocates maintain that being free from overzealous state scrutiny is a fundamental right—and that privacy is essential to the development of the self and the formation of intimate relationships—their opponents routinely point to the dangers of serious crime and terrorism as justifications for greater state surveillance powers.

Perhaps one of the most interesting things about the privacy versus security debate is how little it has changed over the last decade. Despite predictions that an eventual "privacy apocalypse" would shift public opinion and wake people up to the dangers of the surveillance state, there is no indication that recent revelations—such as those exposed by Edward Snowden—have significantly altered the way in which we talk about the conflict between privacy and security. In part, this can be explained by the spread of social media, the growing popularity of online shopping, and the importance of the Internet to both public and private sector service delivery. As we spend more and more time online—and inevitably disclose more information about ourselves to both government and business—our expectations of privacy gradually diminish. As a consequence, demands from the state for more personal information and expanded surveillance powers appear less egregious, if only because we already have less privacy to surrender.

But there are other reasons why the privacy-versus-security debate seems to have reached an impasse or, more pessimistically, why privacy advocates are losing. Because the public debate is inevitably framed in terms of costs, those in favor of privacy often find themselves having to explain abstract notions of personal autonomy and loss of dignity—something that is difficult to do in a newspaper article or radio interview—to an audience increasingly accustomed to giving up privacy in exchange for greater convenience or access to services. In contrast, those in favor of greater surveillance powers can point to tangible past tragedies (like 9/11) while also talking about current and future threats such as a global pandemic. Simply from a rhetorical perspective, privacy advocates have the more difficult side of the debate.

Going further, however, it is clear that this imbalance is not simply about language and the limitations of public debate. It is also about the way in which privacy has been almost exclusively presented as an individual right. Although this is understandable—especially given that the law in many countries protects privacy as a personal interest—this approach has led to a number of unfortunate consequences. First, it exacerbates the tendency to frame arguments about privacy and security in terms of a conflict between the individual and society. More bluntly, it can result in those in favor of robust privacy laws being portrayed as overly self-interested when faced with calls from the police for greater surveillance powers. In democratic states where people generally regard the government as benign and see public agencies—such as the police and security services—as subject to the rule of law, the argument that less personal privacy is a small price to pay for greater collective security is difficult to counter.

Second, the focus on privacy as an individual right exacerbates the tendency to overlook the public value of privacy. The most apparent manifestation of this public aspect of privacy can be found in its role in the promotion and protection of more obviously political rights. Without privacy, many of the rights that are essential to the proper functioning of a democratic society—such as freedom of association and freedom of religion—would be considerably weaker. It is, for example, difficult to enjoy freedom of association if the state is able to know who you meet with, as well as where and when. Similarly, being forced to disclose information about political affiliations or religious beliefs can have a significant chilling effect on the willingness of individuals to engage in political activity or to practice their religion. Despite the fact that a number of prominent privacy scholars have sought to draw attention to this point,[15] privacy's supporting role in relation to these rights is often ignored, with the result that privacy is undervalued in terms of its social and political value significance.

There is, however, another aspect of the public value of privacy that is routinely ignored by both sides of the privacy-versus-security debate. Privacy, insofar as it represents a claim against the state, also plays a crucial role in defining the fundamental relationship between the public and government. By establishing limits on what the state can legitimately know about us—or what information government agencies can require us to divulge in exchange for protection or public services—privacy helps to mark the boundaries of state power. When asserted against the state, privacy can help to establish and demarcate aspects of life that the state has no business interfering in. In this respect, privacy is valuable because it serves as a brake on the expansive tendencies of the state.

In this chapter, my aim is to explore this neglected aspect of the public dimension of privacy. In particular, I hope to revive interest in what Jed Rubenfeld referred to in 1989 as the "anti-totalitarian theory of privacy."[16] Although rarely invoked in contemporary discussions about the justifications for privacy, Rubenfeld's argument that privacy and anonymity serve as vital protections against the totalizing instincts of the authoritarian state is perhaps more salient now that it has ever been. This is not so much because surveillance technologies are more readily available to governments in one-party states or corrupt regimes but in fact because populist and increasingly authoritarian politics have emerged in well-established liberal democracies like the United States and the United Kingdom. As Kevin Haggerty recently noted:

> While lawyers typically focus on privacy violations—and I unreservedly share those concerns—I also worry about the broader political and social implications of a decline in the reality of privacy, as it is, in part, this ability to remain unknown, untracked, and unseen that insulates people from the extremes of institutional manipulation, coercion, and repression. . . . The loss of privacy is matched by a corresponding and unprecedented intensification in institutions' informational capacities. Such a situation carries with it stark prospects for repressive forms of control, as surveillance is a key power resource in all forms of totalitarianism.[17]

In this chapter, I argue that we need to better protect privacy as a public value if we are to ensure that the growing use of surveillance and increased reliance on information technologies do not result in a steady slide into authoritarianism. In particular, this chapter argues that the law needs to expand its existing approach to privacy with a view to creating physical and virtual spaces where political discourse can thrive, and legitimate forms of resistance—such as public and online protests—can be organized and carried out. At the heart of this argument is a claim that the law needs to take ideas of anonymity seriously and to acknowledge that a key aspect of privacy in the twenty-first century is the ability to undertake certain activities—especially in the political sphere—without being named or otherwise identified by the state. As difficult as it may be to conceive of law's role in this, in order to promote and protect democratic values, the law may need to help us become invisible to the state and—in very limited yet important ways—place ourselves outside of its gaze and, by implication, outside of its reach.

The Public Value of Privacy

Privacy prevents interference, pressures to conform, ridicule, punishment, unfavorable decisions, and other forms of hostile reaction. To the extent that privacy does this, it functions to promote liberty of action, removing the unpleasant consequences of certain actions and thus increasing the liberty to perform them.

—Ruth Gavison, "Privacy and the Limits of Law"

Privacy is notoriously difficult to define. In the opening paragraph of his 2002 article "Conceptualising Privacy," Daniel J. Solove observed, "Time and time again philosophers, legal theorists, and jurists have lamented the great difficulty in reaching a satisfying conception of privacy."[18] Over the course of the next two paragraphs, he lists a series of these laments from some of the most prominent privacy scholars of the previous fifty years. Perhaps the most notable comes from Alan Westin, whose early work on defining privacy remains essential for anyone working in the field: "Few values so fundamental to society as privacy have been left so undefined in social theory."[19] Although a great deal has been written about privacy since Westin first made this observation, it remains remain largely true today. Privacy continues to be a concept that, despite its apparent familiarity, remains stubbornly resistant to efforts to pin it down or find a consensus about why we regard it as valuable. As Solove went on to note, this problem of definition and justification is not just a matter of concern to privacy scholars and advocates. For lawmakers and judges, it is difficult to resolve apparent conflicts between the right to privacy and the state's pursuit of security if they are unable to clearly define the scope of the right or what is lost when it is infringed or curtailed.

> The difficulty in articulating what privacy is and why it is important has often made privacy law ineffective and blind to the larger purposes for which it must serve. . . . Judges, politicians, and scholars have often failed to adequately conceptualize the problems that privacy law is asked to redress. Privacy problems are often not well articulated, and as a result, we frequently do not have a compelling account of what is at stake when privacy is threatened and what precisely the law must do to solve these problems.[20]

For Solove, the problem of definition comes from the recurring desire on the part of scholars and lawmakers to isolate some "essential characteristic" that unites different accounts of privacy. After surveying various definitions and theories—all of which claim, to varying degrees, to have identified the core values that lie at the heart of our general concern with privacy—Solove went on to suggest that privacy is instead best understood as a series of "family resemblances." Drawing explicitly on Wittgenstein and applying what he described as a "pragmatic approach to reconceptualising privacy," Solove argued that we should take a more contextual view of privacy. That is, we should examine those areas and aspects of life where we most commonly encounter privacy claims—"family," "body," "home," etc.—and seek to understand the underlying value of privacy in each specific context. In doing this, the project of understanding privacy becomes less one of definition and delineation, and more one of mapping:

If we no longer look for the essence of privacy, then to understand the "complicated network of similarities overlapping and criss-crossing," we should focus more concretely on the various forms of privacy and recognize their similarities and differences. We should act as cartographers, mapping out the terrain of privacy by examining specific problematic situations rather than trying to fit each situation into a rigid predefined category.[21]

Although there are many reasons to favor Solove's approach, the suggestion that we should avoid looking for "common denominators" when talking about privacy and focus instead on "family resemblances" provides a particularly useful starting point for any discussion about the public and political nature of privacy. If we take a step back from the various accounts of privacy surveyed by Solove, certain resemblances become clear. First, it is apparent that every theory of privacy, regardless of how it answers the question of why privacy is valuable, accepts that it must have some public dimension: that is, privacy claims must be capable of being claims against the state (and not just against private individuals or organizations). In many respects, this is an obvious point. In countries like Canada, Germany, and the United States, the reason privacy has been conferred the status of a right is because it is at its core seen as a claim of noninterference against the state.[22] Although the rationales behind this call for non-interference might differ—being variously grounded in appeals to secrecy, limited access to the self, informational self-determination, the value of intimacy, or identity—the fact remains that privacy is understood as a right that has "vertical" effect.

Second, the search for family resemblances also reveals that most if not all accounts of privacy acknowledge that privacy has some social or political dimension. As Priscilla Regan has observed, although debates about privacy are frequently couched in terms of individual interests, taken together these individual privacy interests also give rise to goods that are enjoyed by all. In this regard, Regan has argued, privacy must be understood as both an individual good and a social good:

> Most privacy scholars emphasize that the individual is better off if privacy exists; I argue that society is better off as well when privacy exists. I maintain that privacy serves not just individual interests, but common, public, and collective purposes. If privacy became less important to one individual in one particular context, or even to several individuals in several contexts, it would still be important as a value because it serves other crucial functions beyond those that it performs for a particular individual.[23]

Of the three forms of social good that Regan has identified—the common, the public, and the collective—it is the second that is perhaps most
immediately relevant to discussions about privacy, democracy, and the
limits of the state. For Regan, the public aspect of privacy is rooted in the
fact that a degree of privacy (and in some cases anonymity) is necessary
for the proper functioning of (liberal) democratic government. As Helen
Nissenbaum has noted, Regan's insight here is valuable because it draws
our attention to the role privacy plays in supporting democratic systems
of government:

> First, [privacy] is constitutive of the rights of anonymous speech and freedom
> of association, and is implicated in the institution of the secret ballot. Second,
> privacy shields individuals against over-intrusive agents of government, par-
> ticularly in spheres of life widely considered out of bounds. . . . Third, privacy
> allows people to separate and place in the background aspects of their private
> lives that generally distinguish them from others [with that result that] when
> people come together in a public realm, they are able to place in the foreground
> what they have in common with their fellow citizens.[24]

Looked at in this way, it becomes clear that privacy is inherently political.
Without privacy and the subsequent possibility of engaging anonymously
with key political processes like elections, it is almost impossible to imagine
how citizens could be full participants in a liberal democratic system of
government that values freedom and takes enfranchisement seriously.

Returning to the question of "family resemblances," it is important to
note that while competing theories of privacy may provide differing accounts
of what it is that privacy seeks to protect (or the nature of its underlying
value), most explicitly recognize (or at least implicitly acknowledge) that
privacy is necessary to the exercise of fundamental rights such as freedom
of expression, freedom of religion, and freedom of association. Put another
way, although we may argue about why privacy deserves to be regarded as
a "first-order" right, there is little disagreement over its importance as a
"second-order right." For example, where individuals and groups are deprived
of the ability to keep their communications private—or prevented from
meeting without being identified and monitored by the state—it is difficult
for them to fully enjoy the social or political dimensions of these rights.[25]

Although Regan has made similar arguments when speaking about
aspects of both the "common" and "public" values of privacy, the point
here is a slightly different one. For Regan, the key is to recognize that the
personal autonomy and freedom of speech, religion, and association made

possible through the exercise of individual privacy rights give rise to social and political outcomes that are good for everyone in a liberal democracy. In this respect, her argument is that privacy is good not just for our individual well-being but also for the general health of democratic structures and institutions of government. However, I would like to suggest that Regan's observation needs to be taken a step further. Our individual and collective enjoyment of the benefits that come from the exercise of first-order rights like freedom of speech, freedom of religion, and freedom of association is not simply enhanced by the existence of privacy; rather, these benefits are made possible by privacy. Without some idea of privacy—understood as a limit placed on the ability of the state to identify and observe us—it is not just difficult to enjoy these rights; it is hard to describe them as rights at all.

Revisiting the Anti-Totalitarian Theory of Privacy

Assuming we accept this mapping of the broad landscape of the right to privacy—and how the various "family resemblances" identified in this chapter have been framed—where does that leave us in terms of our discussion about the public value of privacy and in particular the role it can play in helping to constrain the state's authoritarian instincts? How do these observations about privacy move us closer to accepting—and acting on—the claim that there should be spaces where we can be invisible to the state? On the one hand, it can be argued that once we accept that privacy has a public dimension—and that we need privacy in order to enjoy rights such as freedom of expression and freedom of association—then it becomes easier to accept the claim that privacy has an inescapable political dimension. As has already been noted, one of the challenges for privacy advocates is trying to explain why privacy matters, especially in the face of arguments about the dangers of crime and terrorism. By drawing attention to the political aspects of privacy, we shift the focus of the privacy-versus-security debate: from the supposed tension between individual rights and public safety to the question of how to reconcile competing public goods (privacy and security). This shift is important because it helps to ensure that the costs of state surveillance—in terms of reduced privacy—are framed as costs not only to individuals or specific groups. Instead, increased state surveillance is shown to have a social cost insofar as it threatens political freedom and has the potential to make the exercise of key democratic rights—such as freedom of expression and association—more difficult.

Going further, by acknowledging that privacy is necessary to the proper functioning of a liberal, democratic state, we also make explicit the idea that—in certain contexts at least—we need privacy to protect us from a potentially authoritarian state. This is an important point and one that was made almost three decades ago by Jed Rubenfeld in his 1989 article "The Right to Privacy." According to Rubenfeld, theories of privacy based on the idea of personhood are fundamentally misguided, if only because the right to define oneself is irretrievably incoherent. Given this, Rubenfeld argued that the only convincing basis for recognizing a right to privacy is the need to protect ourselves from the overwhelming, normalizing power of the state:

> We are all so powerfully influenced by the institutions within which we are raised that it is probably impossible, both psychologically and epistemologically, to speak of defining one's own identity. The point is not to save for the individual an abstract and chimerical right of defining himself; the point is to prevent the state from taking over, or taking undue advantage of, those processes by which individuals are defined in order to produce overly standardized, functional citizens.[26]

According to Rubenfeld's "anti-totalitarian theory," privacy is valuable because it places important limits on the state—limits that are vital in order to restrain the authoritarian impulses of those in power and prevent the "normalization" of society:

> The anti-totalitarian right to privacy, it might be said, prevents the state from imposing on individuals a defined identity. . . . The danger [to be avoided] is a particular kind of creeping totalitarianism, an unarmed *occupation* of individuals' lives. That is the danger of which Foucault as well as the right to privacy is warning us: a society standardized and normalized, in which lives are too substantially or too rigidly directed. That is the threat posed by state power in our century.[27]

Although subsequent writers such as Solove have drawn attention to flaws in Rubenfeld's rejection of the "personhood idea of privacy,"[28] his efforts to put the state at the center of his analysis are worth revisiting, particularly in light of the dramatic spread and intensification in state surveillance over the past three decades. While Solove may be right when he suggests that this "conception of privacy collapses into a vague right to be let alone," Rubenfeld was nonetheless right to draw our attention to the inherently political nature of privacy and the totalitarian urges of the state:

> The right to privacy is a political doctrine. It does not exist because individuals have a sphere of "private" life with which the state has nothing to do. The state has everything to do with our private life; and the freedom that privacy

protects equally extends, as we have seen, into "public" as well as "private" matters. The right to privacy exists because democracy must impose limits on the extent of control and direction that the state exercises over the day-to-day conduct of individual lives.[29]

Even if the anti-totalitarian theory does not provide a fully satisfying conception of privacy, Rubenfeld's key insight—that we should put the state at the center of our discussion about what is at stake when we talk about privacy—is particularly apposite in the current climate. More crucially, by drawing attention to the fact that the ability to identify individuals— and through that process of identification to impose certain identities on them—is a crucial component in the exercise of authoritarian power, Rubenfeld provides reason for revisiting a neglected but important aspect of privacy, namely anonymity.

While privacy, as Solove argues, can be understood only as a series of interconnected ideas and defined only according to specific contexts, anonymity is a narrower claim. A demand for anonymity is a demand to be free from identification, and while this notion will be discussed further below, it is clear that in both principle and practice, the link between this claim and the danger that identification poses to the exercise of key political rights is a close one. We may disagree about why privacy deserves to be protected, but most of us would accept that there are very good reasons to keep certain aspects of our political lives—such as our voting choices and political affiliations—from the state. The real question, however, is whether we need to extend the more common and basic conception of anonymity and recognize that there are other contexts in which it may be necessary to shield ourselves from identification by the state. As Nissenbaum has noted:

> If, as a society, we agree that what is importantly at stake in anonymity is the capacity to be unreachable in certain situations, then we must secure the means to achieve this. This will include a dramatic reversal of current trends in surveillance, as well as a relentless monitoring of the integrity of systems of opaque identifiers. Without at least these measures, even if we nominally secure a right to anonymity through norms and regulations, we will not have secured what is at stake in anonymity in a computerized world.[30]

This is an important point. As has already been noted at the beginning of this chapter, the last thirty years have seen a massive increase in both the reach and intensity of state-sponsored surveillance in democratic countries like the United States and the United Kingdom. As a consequence, many of the concerns expressed by Rubenfeld in 1989 about the totalizing effects of state scrutiny are far more pressing than they were when his article was first published. Going further, in light of changes in the political landscape in the United States since the 2016 presidential election and the reemergence of the extreme right and nationalist politics in countries across Europe, there are good reasons we may wish to shore up the protection of fundamental democratic rights like freedom of speech, freedom of religion, and freedom of association through a greater commitment to anonymity.

Embracing Anonymity

> Anonymity is a shield from the tyranny of the majority. It thus exemplifies the purpose behind the Bill of Rights, and of the First Amendment in particular: to protect unpopular individuals from retaliation—and their ideas from suppression—at the hand of an intolerant society. The right to remain anonymous may be abused when it shields fraudulent conduct. But political speech by its nature will sometimes have 357 consequences, and, in general, our society accords greater weight to the value of free speech than to the dangers of its misuse.
>
> —Associate Justice John Paul Stevens, *McIntyre v. Ohio Elections Commission*

> Anonymity facilitates the flow of information and communication on public issues, safeguards personal information, and lends voice to individual speakers who might otherwise be silenced by fear of retribution.
>
> —Ian Kerr and Alex Cameron, *Contours of Privacy*

Derived from the Greek *anonymia* (ἀνωνυμία), the word *anonymity* originally referred to something that was nameless or "un-named."[31] As various writers have observed, however, anonymity is now best understood as a claim against identifiability. According to K. A. Wallace, for example, "Anonymity is a kind of relation between an anonymous person and others, where the former is known only through a trait or traits that are not coordinatable with other traits such as to enable identification of the person as a whole."[32]

Although privacy has been the subject of a great deal of academic and judicial interest, surprising little has been written about the relationship between privacy and anonymity. Instead, where anonymity has been the focus of attention, it has often been in the context of discussions about freedom of expression. In the United States, for example, anonymity has been recognized by the Supreme Court as a "subcategory of the constitutional guarantee of silence" and is protected by the First Amendment.[33] As Martin Redish has observed, US courts have held that protecting anonymity is, in certain contexts, an essential precursor to protecting political speech. In this regard, anonymity is by its very nature inherently political:

> In one important sense, of course, the right of anonymity qualitatively differs from the right not to speak. While the latter could be construed to apply to a generic right to keep silent, the right of anonymity represents an expressive hybrid. It applies when and only when one first chooses to speak, write, or associate for political purposes. The right of anonymity, then, is a selective form of expressive silence: it is only when the speaker first affirmatively chooses to speak that this form of silence comes into play. Nor are the theoretical underpinnings of the right of anonymity identical to the rationale for a generic right to remain silent. While the latter is grounded primarily in the desire to avoid humiliation, demoralization, and cognitive dissonance, the former is designed to avoid chilling the speaker's willingness to contribute fully and frankly to public discourse without fear of retribution from either government or private power centers.[34]

While this particular focus on freedom of expression helps to explain why so little has been written about the relationship between privacy and anonymity, there are other reasons for this lacuna. On the one hand, there has been a tendency on the part of both scholars and lawmakers to conflate privacy and anonymity, and to see anonymity as a *means* to protecting privacy rather than as a distinct right or end in itself.[35] As Skopek has noted, however, this conflation obscures the fact that anonymity has a different albeit deeply related function from privacy:

> Although both anonymity and privacy prevent others from gaining access to a piece of personal information, they do so in opposite ways: privacy involves hiding the information, whereas anonymity involves hiding what makes it personal. The second point is about their formal relationship. Anonymity and privacy have the same causal origin and thus are flip sides of each other. They describe opposite sides of a single underlying event.[36]

Another reason why anonymity may not have received the attention it deserves—in debates about privacy at least—is because we are simply suspicious of individuals who wish to be anonymous. Why would anyone want to be anonymous unless they are doing something that is either wrong or would likely draw criticism? Anonymity conjures up images of people acting in the shadows, giving them a license to engage in conduct and express views that are at best undesirable and at worst criminal. As Nick Harkaway has observed, this fear of anonymity is not entirely unfounded. As anyone familiar with Internet forums or social media can attest to, online anonymity can lead individuals to behave in ways that are far outside accepted norms of behavior:

> Anonymised and disconnected (in the face-to-face sense) from the people from whom they interact, internet users can become spiteful and splenetic to a degree that would never be permitted in a physical social context, or, if it were, it would be profoundly uncomfortable and might devolve into violence. This can be seen as another aspect of the crowd phenomenon: self-reinforcing certainties unchecked by social brakes derived from actual presence.[37]

Regardless of why anonymity has not received more attention, there are good reasons to take it seriously, even if we are inclined toward thinking of privacy as an exclusively individual right. While it is true that freedom from identification can make it difficult to hold people accountable for views and behaviors that we find offensive, it is this same absence of accountability that enables people to speak freely, explore difficult ideas, and give expression to aspects of their identity in ways that would be impossible were they to be personally identified.[38] In addition, the ability to be anonymous in an increasingly connected world enables us to be alone, even though we may— both physically and virtually—be surrounded by others. Anonymity offers us space: both to pursue the development of the self and to interact with others without having to reveal ourselves or "entertain their subjectivity."[39]

Leaving aside questions about the individual value of anonymity, it is also clear that the ability to be free from identification has a deeply social and political dimension. As Regan, Nisenbaum, and others have pointed out, for example, being able to cast one's vote anonymously is central to the proper functioning of any democratic system of government. Similarly, the ability of individuals to express their political views in the company of others and to form political associations free from unwanted scrutiny by government is an essential feature of the liberal state. This is a point that has been made by Michael Froomkin, who has argued that "communicative anonymity is a core part of freedom in a democratic state and a critical tool for those who seek freedom from nondemocratic states."[40]

Like privacy, then, anonymity has a public dimension that extends well beyond whatever benefit it confers on the individual. Returning to Regan's discussion of the public value of privacy above, anonymity "shields individuals against over-intrusive agents of government" and "allows people to separate and place in the background aspects of their private lives." Equally, anonymity is essential to the proper functioning of key political rights like freedom of expression and freedom of association. In this regard, it serves as a second-order right, ensuring that the conditions necessary for the exercise of political rights are present and that the state's desire to know about us does not lead to the suppression of those rights.

On its face, this may sound circular: when we talk of anonymity here, are we really just speaking about another manifestation of privacy? But it is important to distinguish between the two for a number of reasons. First, concerns about privacy tend to focus on access to information either "about" individuals or somehow "belonging" to them. This is true regardless of whether we are focused on the individual or the public value of privacy. In contrast, anonymity is concerned with identification and the problem of "being seen." Taken together, they play complementary but different roles in facilitating the exercise of other substantive rights like freedom of speech and freedom of association.

To make this point clearer, imagine a society in which people are free to form political associations and meet without interference from the state. In addition, the content of any communication between the members of such associations is protected as private and not subject to government scrutiny. However, there are no restrictions on the state when it comes to identifying members of the association or knowing where and when the group meets. In such an example, we would say that the members have a degree of privacy but no anonymity. In such a case, we might imagine that members of the group may be reluctant to meet, not just because they worry that the state will draw inferences about them based on their associates but because of the ability of the state to connect their very membership with other pieces of information legitimately within its possession (such as tax returns, immigration and travel records, and birth certificates). As Froomkin has noted, depending on the circumstances, the move from identification to suppression can be a relatively easy one:

> In many countries, the power to prevent anonymity, to force identification, and to gather traffic data for analysis will be used to stamp out dissidents. We should admit that sometimes those dissidents may be terrorists; technology can empower very bad people as well as very good ones. But that is also [the] point: sometimes the very bad people are in power, and the people against whom they will use technologies of identification are the human rights activists, the democratic and nonviolent protestors, and the Twitter users planning demonstrations. And after the technologies of identification will come the technologies of retaliation.[41]

If this example seems too far from home, one only needs to recall the Justice Department's recent demand for the names of every Internet user who visited an "anti-Trump" website in the days leading up to the January

2017 presidential inauguration. Although lawyers for DreamHost, the Internet company that hosted the site, argued that "the search warrant not only aims to identify the political dissidents of the current administration, but attempts to identify and understand what content each of these dissidents viewed on the website," the Court upheld an amended version of the warrant and ordered DreamHost to disclose the requested information.[42] In a similar move, in September 2017 the Justice Department served Facebook with three search warrants, demanding the private account information of potentially thousands of Facebook users in its efforts to identify "anti-administration activists."[43] Returning to Froomkin's recent analysis of the importance of anonymity and the current political climate in the United States:

> If it was not clear enough already, the results of the U.S. 2016 election should re-emphasize the importance of preserving not only the right to speak anonymously but also the practical ability do so. The internet and related communications technologies have shown a great potential to empower end-users, but also to empower firms and especially governments at the end-users' expense. Governments (and firms) around the world have learned this lesson all too well, and are taking careful, thorough, and often coordinated steps to ensure that they will be among the winners when the bits settle.[44]

So where does this leave us? Given that identification is often accompanied by state-sponsored efforts to suppress political dissent—even in countries that are regarded as liberal democracies—it seems clear that if we accept privacy has value as a public good, then we have to take anonymous speech and anonymous association more seriously too. We need to explicitly recognize that in some contexts, more privacy—in the form of limitations on the ability of the state to collect, use, and share our personal information—may not be enough. What might be needed is more anonymity.

On its face, the call for more anonymity may not seem controversial. But if we turn our minds to what a commitment to anonymity—at least in terms of people's political lives—might look like in practice, it is easy to anticipate the many objections of those who would favor security over privacy. This is because a genuine commitment to anonymity means being willing to say that sometimes there are things the state cannot know, and sometimes there are places where the state cannot go, even if this means limiting the investigative powers of law enforcement and potentially making society less safe. In light of the sophistication of modern surveillance techniques and the information that can be gained from facial recognition software, data matching, metadata, and ISP addresses, it is almost impossible to monitor individuals or their online activity without exposing them to subsequent identification. As a consequence, any form of surveillance inevitably raises the prospect of individuals being identified and thereby being made vulnerable to the state. Given that the mere prospect of being identified can have a chilling effect on the exercise of free speech and freedom of association, there is no middle ground: we must be willing to guarantee anonymity in some contexts if we are to maintain a commitment to free political discourse and create spaces for the genuine fostering of dissent.

In practical terms, this commitment to the public value of anonymity would lead to substantial limits being placed on the surveillance power of the state. First, it would require us to create—or at least allow for the possibility of—dedicated physical and virtual spaces for political discourse, spaces that are free from any form of government oversight. Returning to the DreamHost example, this would mean allowing websites that promote political speech and dissent to guarantee their users absolute anonymity. Equally, it would mean placing physical limits on the surveillance powers of the police in order to ensure that people are able to meet in person and take part in public protests without any fear of being "seen" by the state. The use of CCTV and other forms of electronic monitoring would, for example, be forbidden at political demonstrations. Finally, a commitment to anonymity would also require us to engage in a serious conversation about information-sharing between the private sector and the state, in particular about whether social media platforms should be able to provide users with guarantees of anonymity (and to lawfully resist efforts by the state to identify those users).

Although these proposals are general in nature, they are underpinned by two basic assumptions about the public value of privacy and the importance of anonymity. First, they take as a given that identification constitutes a significant infringement on privacy and has serious implications for the exercise of political rights. Second and perhaps more controversially, they assume that even liberal states have authoritarian tendencies that, if left unchecked, constitute a potential threat to the political freedom and well-being of its own citizens. It is this second point that, following on from Rubenfeld, may be the hardest to swallow for those accustomed to viewing the state as essentially benign. Going further, creating spaces of anonymity may seem like an overreaction, as it would impose real limits on the ability of the police and other law enforcement agencies to combat crime and terrorism. However, given just how extensive the surveillance powers of the modern state have become, it can be argued that we underestimate the threat posed by the state—even the liberal, democratic state—at our peril. As Kevin Haggerty has observed:

The loss of privacy [has been] matched by a corresponding and unprecedented intensification in institutions' informational capacities. Such a situation carries with it stark prospects for repressive forms of control, as surveillance is a key power resource in all forms of totalitarianism. . . . History has demonstrated how such repression becomes easier when privacy is reduced by even small increases in organizational information processing capabilities. Such surveillance resources have never been more powerful or accessible.[45]

It is also important to note that none of the above arguments prevent the state from placing reasonable restrictions on the circumstances under which people are able to exercise their rights to freedom of expression and freedom of association anonymously. The fact that a website simply claims to be a space for free political discourse should not, for example, be enough to exempt it from government oversight or surveillance. Instead, it is possible to imagine a system similar to the granting of charitable status being established in order to distinguish between sites that are genuinely committed to promoting free speech and those that simply make this claim to put themselves out of the reach of the state. Similarly, sites that create a platform for hate speech or online harassment should not be able to guarantee their users anonymity. While distinguishing between genuinely political spaces and "everything else" is likely to present a significant regulatory challenge, even an imperfect process is preferable to none at all, especially if it results in the creation of at least some spaces where political dissent can exist outside of the reach of the state.

Before leaving this discussion, it is worth reflecting on the impact these proposals may have on the ability of law enforcement and the intelligence services to prevent crime and combat terrorism. There is little doubt that, regardless of how carefully we construct a system that creates spaces of anonymous political discourse, such spaces will inevitably attract criminals and potential terrorists keen to be free from state surveillance. While the existence of anonymous spaces may make it more difficult for law enforcement to monitor the activities of such individuals, given the various investigative and surveillance tools at the disposal of the modern state, it does not necessarily follow that allowing for anonymity in a limited range of contexts will make us less safe. Instead, a commitment to anonymity for the purpose of protecting political speech and checking the authoritarian instincts of the state may require law enforcement to reexamine the ways in which it uses various surveillance technologies in combination with other forms of information gathering. As surveillance technologies have become ever more sophisticated, there has been a tendency on the part of law enforcement agencies to rely on technological solutions to the problem of investigating and preventing serious crime and acts of terrorism, often at the expense of other, more traditional approaches (such as community outreach, human informers, and physical searches). Depriving the police of the ability to engage in electronic surveillance in some online or physical contexts does not mean that they are powerless to identify people in their efforts to combat crime—any more than the need for a search warrant to enter a person's home means that a police investigation has to stop when confronted with a closed door.

As a final point, it is worth noting that even if a commitment to greater anonymity means exposing the public to a greater risk of crime—or even terrorism—it may be a price worth paying. The twentieth century is replete with examples of what happens when a previously benign state turns against its own citizenry, and in almost every case the suppression of political dissent was the first step toward the horrors of totalitarianism. Assuming that the sorts of political upheavals that occurred in Europe during the 1930s could not happen in democracies like Australia, Canada, the United Kingdom, or the United States is not only naive; it ignores the fact that those states already have surveillance capacities that the authoritarian governments of countries like the former East Germany could only dream of. As such, the threat posed by state surveillance to freedom of speech, freedom of association, and robust political discourse is very real and needs to be taken seriously. Although there are many challenges associated with anonymity, it has the virtue of being a blunt tool—one that operates only in terms of the binary of known or not known. As such, it offers a means to imposing meaningful limits on the surveillance power of the state and, with those limits, the continuing possibility of free political discourse.

Conclusion

I began this chapter by detailing the many ways in which the state renders us visible. In making the argument that we need to create spaces of anonymity in order to protect key political rights from the chilling effects of state surveillance, I have ended with a confronting proposal: that the state must also be willing to countenance the possibility that we can and sometimes should be invisible.

In a world in which more knowledge is almost blithely assumed to be better than less, the idea that we would carve out spaces in which we cannot know what people are doing or saying is hard to countenance. And while there are clear dangers associated with tolerating anonymous speech and allowing individuals to be unidentifiable in certain contexts, the dangers posed by an increasingly overwhelming architecture of state surveillance, even in supposedly liberal democracies, are even greater. If we are to take the public value of privacy seriously and acknowledge its vital role in the protection of fundamental political rights, then we need to talk more about anonymity. Although Rubenfeld's anti-totalitarian theory may have been overlooked in favor of other, more comprehensive accounts of privacy, his central insight—that we need privacy to restrain the power of the state—has never been more relevant.

Notes

1. Since this chapter was first drafted, the Black Lives Matter (BLM) movement has gained considerable momentum and become a prominent force in American public discourse. The inherent imbalances between the way groups of Americans experience state authority have been brought into stark relief by BLM protests, rallies, and marches, as well as by increasingly mainstream coverage of violent police action. This chapter does not explore the ways in which these recent dynamics may fundamentally shift consensus around the visibility of individuals vis-à-vis the state, but these issues clearly deserve the attention of privacy and surveillance scholars.

2. There is now a wealth of literature on the subject of electronic surveillance, in particular on the social, political, and cultural implications of the phenomenon of state-sponsored mass surveillance. For an overview of some of the key themes and concerns in this literature, see Sean Hier and Joshua Greenberg, eds., *The Surveillance Studies Reader* (Milton Keynes: Open University Press, 2007); Kirstie Ball, Kevin Haggerty, and David Lyon, eds., *Routledge Handbook of Surveillance Studies* (Abingdon: Routledge, 2012); David Lyon, *Surveillance after Snowden* (London: Polity Press, 2015); and Bernard Harcourt, *Exposed: Desire and Disobedience in the Digital Age* (Cambridge, MA: Harvard University Press, 2015).

3. Central to many of the Snowden disclosures is the Five Eyes (FVEY), an intelligence-sharing alliance between Australia, Canada, New Zealand, the United Kingdom, and the United States. All five countries are also parties to the UKUSA Agreement, a multilateral treaty that sets out the terms for joint cooperation and the sharing of information gained through signals intelligence. For a summary of Snowden's key disclosures, see Glenn Greenwald, *No Place to Hide: Edward Snowden, the NSA, and the U.S. Surveillance State* (New York: Metropolitan Books, 2014); Luke Harding, *The Snowden Files: The Inside Story of the World's Most Wanted Man.* (New York: Vintage Books, 2014).

4. For examples of the concerns raised by NGOs about surveillance and its impact on individual privacy and political freedom, see the relevant documents and reports on the websites of the ACLU, Human Rights Watch, and EPIC.

5. James Rule, *Private Lives and Public Surveillance* (London: Allen Lane, 1974), 31.

6. Ipsos Cybersecurity Poll Conducted for Reuters, March 31, 2017, https://www.reuters.com/article/us-usa-cyber-poll-idUSKBN1762TQ.

7. Vian Bakir, Jonathan Cable, Lina Dencik, Arne Hintz, and Andrew McStay, *Public Feeling on Privacy, Security and Surveillance: A Report by DATA-PSST and DCSS*, 2015.

8. Kirstie Ball, David Lyon, David Murakami Wood, Clive Norris, and Charles Raab, *A Report on the Surveillance Society* (London: Office of the Information Commissioner, 2006), 1.

9. British Security Industry Association, *The Picture Is Not Clear: How Many CCTV Surveillance Cameras in the UK*, form no. 195, issue 1 (2010).

10. According to a report by Next City in June 2017, the New York Police Department (NYPD) maintains approximately 2,000 public area CCTV cameras and has access to an additional 7,000 cameras in public housing and a further 4,000 in the subway system. The report also notes that the NYPD can readily access 4,000 private security cameras across the five boroughs. See https://nextcity.org/daily/entry/new-york-surveillance-cameras-police-safety.

11. As Cory Doctorow observed in the *Guardian* in 2015, "Spying is cheap." See Cory Doctorow, "Technology Should Be Used to Create Social Mobility—Not to Spy on Citizens," *Guardian*, March 10, 2015.

12. See https://www.aclu.org/issues/national-security/privacy-and-surveillance/nsa-surveillance. In January 2018, Congress approved an extension to Section 702 of FISA, extending the power of law enforcement agencies like the FBI to intercept and record the communications—such as emails and phone calls—of non-US citizens outside the United States without a warrant.

13. One of the most explicit examples of this can be found in Article 8 of the *European Convention on Human Rights*. Although Article 8 establishes that "everyone has the right to respect for his private and family life, his home and his correspondence," it also substantially qualifies the right, allowing states to limit the scope of the right "in accordance with the law and is necessary in a democratic society in the interests of national security, public safety or the economic well-being of the country, for the prevention of disorder or crime, for the protection of health or morals, or for the protection of the rights and freedoms of others." See also the discussion of how Canadian courts have approached the issue of privacy at the border in Benjamin Goold, "Privacy Rights at the Canadian Border: Judicial Assumptions and the Limits of the *Charter*," in Catherine Dauvergne, ed., *Research Handbook on the Law and Politics of Migration* (Cheltenham: Edward Elgar, 2021).

14. Michael Froomkin, "Lessons Learned Too Well: Anonymity in a Time of Surveillance," *Arizona Law Review* 59, no. 1 (2017): 95.

15. See in particular Priscilla Regan and Helen Nissenbaum, both of whom are discussed in the next section.

16. Jed Rubenfeld, "The Right to Privacy," *Harvard Law Review* 102 (1989): 737.

17. Kevin Haggerty, "What's Wrong with Privacy Protections? Provocations from a Fifth Columnist," in Austin Sarat, *A World without Privacy: What Law Can and Should Do?* (Cambridge: Cambridge University Press, 2014), 198.

18. Daniel Solove, "Conceptualising Privacy," *California Law Review* 90, no. 4 (2002): 1087. Among the authors he cites here are Ruth Gavison, Arthur R. Miller, Julie C. Innes, and Tom Gerety. Perhaps the most disheartening of the quotes comes from Robert Post, who states that "privacy is a value so complex, so entangled in competing and contradictory dimensions, so engorged with various and distinct meanings, that I sometimes despair whether it can be usefully addressed at all." See Robert C. Post, "Three Concepts of Privacy," *Georgetown Law Journal* (2001): 2087 (quoted in Solove, 1089).

19. Alan Westin, *Privacy and Freedom* (New York: Atheneum, 1967), 7.

20. Solove, "Conceptualising Privacy," 1090. According to Solove (1094), attempts to conceptualize privacy can be divided into six overlapping (yet distinctive) accounts: (1) the right to be let alone; (2) limited access to the self; (3) secrecy; (4) control of

personal information; (5) personhood; and (6) intimacy. In contrast, Judith Thompson has argued that privacy may defy definition: "Nobody seems to have any very clear idea what the right to privacy is. We are confronted with a cluster of rights—a cluster with disputed boundaries—such that most people think that to violate at least any of the rights in the core of the cluster is to violate the right to privacy; but what have they in common other than their being rights such that to violate them is to violate the right to privacy? To violate these rights is to not let someone alone? To violate these rights is to visit indignity on someone?" See Judith Thompson, "The Right to Privacy," *Philosophy & Public Affairs* 4, no. 4 (1975): 295–314, quote at 313.

21. Solove, 1126.

22. It is worth noting that while many jurisdictions allow individuals to sue organizations or other individuals for breaches of privacy, such actions are usually grounded in specific privacy legislation.

23. Pricilla Regan, *Legislating Privacy: Technology, Social Values, and Public Policy* (Chapel Hill: University of North Carolina Press, 1995), 221.

24. Helen Nissenbaum, *Privacy in Context: Technology, Policy, and the Integrity of Social Life* (Stanford, CA: Stanford University Press, 2010), 86–87.

25. This is a point I have made in earlier work on the political value of privacy: Benjamin Goold, "Surveillance and the Political Value of Privacy," *Amsterdam Law Forum* 1, no. 4 (2009): 3–6. As I noted, "By focusing on the political rather than the individual dimension of privacy, we not only free ourselves from complex discussions of individual autonomy and dignity, but also ensure that the relationship between the individual and the state remains at the heart of any debate about privacy and surveillance."

26. Rubenfeld, "The Right to Privacy," 794.

27. Rubenfeld, 794.

28. According to Solove: "Although Rubenfeld's critique of the personhood conception is certainly warranted, he fails in his attempt to abandon a personhood conception. If privacy concerns only those exercises of state power that threaten the 'totality of our lives,' then it is difficult to conceive of anything that would be protected. . . . Rubenfeld's critique of personhood forbids him to sketch any conception of identity that the law should protect, for to do so would be to seize from individuals their right to define themselves. By abandoning any attempt to define a conception of identity, Rubenfeld's conception of privacy collapses into a vague right to be let alone." See Solove, "Conceptualising Privacy," 1120. As Solove notes at the end of his critique of Rubenfeld: "Although Rubenfeld is correct that the state cannot be neutral when it becomes involved in one's self-definition, he errs in assuming that he can develop his theory of anti-totalitarianism without an account of personhood."

29. Rubenfeld, 804–5.

30. Helen Nissenbaum, "The Meaning of Anonymity in an Information Age," *Information Society* 15, no. 2 (1999): 144.

31. Kathleen Wallace, "Anonymity," *Ethics and Information Technology* 1, no. 1 (1999): 21–35.

32. Wallace, 21.

33. Martin Redish, "Freedom of Expression, Political Fraud, and the Dilemma of Anonymity," in *Speech and Silence in American Law*, ed. Austin Sarat (Cambridge: Cambridge University Press, 2009), 143.

34. Redish, 143.

35. James M. Skopek, "Reasonable Expectations of Anonymity," *Virginia Law Review* 101, no. 3 (2015): 692. According to Skopek, the US Supreme Court's conclusion that the Fourth Amendment's protections do not apply to information that has been exposed to the public or third parties (including information about our "public movements,

Internet usage, cell phone calls") and that it is open to the police and other law enforcement agencies to collect this information derives from this mistaken conflation of privacy and anonymity. For Skopek, the Fourth Amendment gives rise not only to a "reasonable expectation of privacy" but also to a "reasonable expectation of anonymity."

36. Skopek, 694.
37. Nick Harkaway, *The Blind Giant: How to Survive in the Digital Age* (London: John Murray Publishers, 2012), 158.
38. This point is extremely well captured by Helen Nissenbaum: "[Anonymity] offers a safe way for people to act, transact, and participate without accountability, without others 'getting at' them, tracking them down, or even punishing them. [As such, it] may encourage freedom of thought and expression by promising a possibility to express opinions, and develop arguments, about positions that for fear of reprisal or ridicule they would not or dare not do otherwise. Anonymity may enable people to reach out for help, especially for socially stigmatized problems like domestic violence, fear of HIV or other sexually transmitted infection, emotional problems, suicidal thoughts." See Nissenbaum, "The Meaning of Anonymity in an Information Age," 142.
39. Janna Malamud Smith, *Private Matters: In Defense of the Personal Life* (Emeryville, CA: Seal Press, 2003), 45. As Malamud Smith notes, "Anonymity in an urban setting is in some ways equivalent to solitude in nature. Like solitude, anonymity offers space."
40. Froomkin, "Lessons Learned Too Well," 145. As Froomkin goes on to note (145): "The rapid deployment of profiling and surveillance technologies in both the public and private sectors only increases the importance of preserving the ability to be anonymous: without it, every utterance, every purchase, and every computer-mediated interaction risks becoming part of one's dossier. There are profound differences between dossiers kept by marketers, by benign governments, and by malign ones, but in every case without the ability to be anonymous at least sometimes, life becomes a continuous experience of being watched and recorded. . . . Anonymity is the escape hatch."
41. Froomkin, 146.
42. See Charlie Savage, "Justice Dept. Demands Data on Visitors to Anti-Trump Website, Sparking Fight," *New York Times*, August 15, 2017; Morgan Chalfant, "DreamHost to Appeal Ruling on DOJ Request for Data on Anti-Trump Site," *The Hill*, September 5, 2017.
43. See Jessica Schneider, "DOJ Demands Facebook Information from 'Anti-Administration Activists,'" *CNN*, September 29, 2017, https://www.cnn.com/2017/09/28/politics/facebook-anti-administration-activists/index.html.
44. Froomkin, "Lessons Learned Too Well," 159. This is a point that is also made in the Intervener Submission by the Canadian Internet Policy and Public Interest Clinic (CIPPIC) in *BMG Canada Inc v. Doe* 2004 FC 488, quoted in Ian Kerr and Alex Cameron, "Scoping Anonymity in Cases of Compelled Disclosure of Identity: Lessons from *BMG v. Doe*," 3–30. As noted on page 5: "The Internet provides an unprecedented forum for freedom of expression and democracy. The ability to engage in anonymous communications adds significantly to the Internet's value as a forum for free expression. Anonymity permits speakers to communicate unpopular or unconventional ideas without fear of retaliation, harassment, or discrimination. It allows people to explore unconventional ideas and to pursue research on sensitive personal topics without fear of embarrassment."
45. Haggerty, "What's Wrong with Privacy Protections?," 199.

Contributors

LAWRENCE DOUGLAS, Department of Law, Jurisprudence and Social Thought, Amherst College

BENJAMIN J. GOOLD, The Peter Allard School of Law, University of British Columbia

TORIN MONAHAN, Department of Communication, The University of North Carolina at Chapel Hill

KELLI MOORE, Department of Media, Culture, and Communication, New York University

EDEN OSUCHA, Department of English, Bates College

JENNIFER PETERSEN, Department of Media Studies, University of Virginia

CARRIE A. RENTSCHLER, Department of Art History and Communication Studies, McGill University

AUSTIN SARAT, Department of Law, Jurisprudence and Social Thought, Amherst College

MARTHA MERRILL UMPHREY, Department of Law, Jurisprudence and Social Thought, Amherst College

Index

Page numbers in *italics* refer to illustrations.

Goldberg, David Theo, 117
Goldsmith, Andrew, 98–99
Goodrich, Peter, 46, 52, 55, 57, 62n41, 63n46
Good Samaritan laws, 69
Goodwin, Charles, 45
Google: facial recognition program of, 21; Google Street View, 140–43, 153n52–53; privacy of data and, 160
Goold, Benjamin, 180n25
Government Communications Headquarters (UK), 137, 137
government institutions. *See* police; United Kingdom; United States
Grant, Oscar, 79, 116, 123–25
Gray, Freddie, 17
Guardian (UK), data on police shootings, 118

El-Hadi, Nehal, 93, 120–22, 125
Haggerty, Kevin, 6, 163, 175–76
Halliday, George, 123
Halpern, Orit, 135–36
Hammerand, Andrew, 142–43
Han, Sora, 43
Haraway, Donna, 15–16n13
Harkaway, Nick, 156, 171
Harris, Victor, 3–5, 15n9, 40n50
Hartman, Saidiya, 42–43, 114
Hayaert, Valérie, 52
Hemmeter, Marianne, 72
Heyes, Cressida, 76
human attention, objectifying. *See* eye-tracking technology
Humphreys, Lee, 78–79

individual rights to privacy, 160, 162
instrumental realism, 98
Internet. *See* social media and Internet use
Intimate Images and Cyber Protection Act of Nova Scotia (2018), 91n83

Jackson, John L., 124–25
Jonsson, Peter, 84–85
journalism and objectivity, 38n18
Joyce, Patrick, 133

Keller, Evelyn Fox, 37–38n13, 38n19
Kerr, Ian, 170
Khan, Majid, 152n40
Killed by Police (database), 118
King, Rodney, 39n47, 45, 62n33, 123–25
KnightSec, 66
Know My Name (Miller), 84
Krahé, Barbara, 68

law and the visible, 1–16; Anti-Terrorism Crime and Security Act (2001, UK), 158; architectural structure as metaphor for portals to law, 52; Canada and legal sanctions against sexual assault, 68, 82; Cyber Safety Act (2013, Nova Scotia, Canada), 82, 91n92; *facies altera* (law's two faces), 53; FISA Amendments Act of 2008 (FAA), 159, 160, 179n12; George Floyd Justice in Policing Act (2020), 99; growth of visual technologies and, 1–3; justice and visibility, 65–66, 69–72, 80, 83–86; justice gaps in criminal justice approach to sexual violence, 68; legal emblemata, 46, 55, 57, 62n41, 63n46; legal spectatorship, 10, 46–47, 57–59; on liability of bystanders, 69, 75–76, 88n24; PATRIOT Act (2001), 158; Protecting Canadians from Online Crime Act (2015), 82; valuing images as evidence and, 3–8; visual technologies and effect on, 9–14; Youth Criminal Justice Act, 90n79. *See also* archives and counter-archives; bodycams and the Black body; objectivity and perspective in visual technology; police; police-civilian encounters recorded by visual technology; privacy; sexual violence and bystander responsibility
Lenehan, Gregory, 80
"Lessons Learned Too Well" (Froomkin), 181n40, 181n44